Dedicated to the future generations
in the hope that their world may be
bright.

# Contents

# Introduction

This book has a dual purpose. First it is intended to show that the study of physics is very relevant to contemporary environmental problems. The second purpose is to acquaint readers with some of the details of these problems. Everyone should have an understanding of what needs to be improved and what can not be done. There is a rational middle ground between the extremes of "there is nothing wrong with the environment; full speed ahead" and "we are about to be obliterated; full speed reverse". If some readers become sufficiently interested in our environment that they choose careers in this area, then this book will be amply justified. Certainly there is a growing need for people with a thorough understanding of, and competence to solve, our environmental problems.

One of the most important inputs for a proper evaluation of our environmental problems is quantitative information. It makes no sense to focus attention on a minor polluter while ignoring a much larger problem. If a given energy source is not capable of supplying a significant amount of power, little effort should be directed towards developing it. The power and energy needed by an automobile must be known and compared to the capabilities of alternative power sources. Quantitative estimates of the water losses in a wet cooling tower are needed. One can not require physically impossible performance from a mass trasportation system. And the list goes on. In almost every subject numerical results or estimates must be used for a rational decision. This is the place where physics is especially relevant. One view of physics is that its goal is to make mathematical models of nature. Quantification is part of the essence of physics. About half of the Units in this book have detailed quantitative considerations, and the rest include some quantitative aspects. One valuable lesson learned in the study of physics is a judgment of what must be included in a model and what can be neglected successfully. This lesson carries over to environmental considerations.

This book is designed to parallel the order of topics found in most introductory physics courses. It assumes that the basic physics has been covered in the standard texts. A suggested guide to what Unit is appropriate with the material of several texts is included at the end of the Introduction. Although some calculus is used in the present book, its use is not essential, as these places can be omitted.

This text could also be used for an environmental course which has physics as a prerequisite. In this case it would be appropriate to consider all of the Units on a given topic together. A suggested arrangement is given later.

There are some problems given at the end of each Unit. In many cases the problems are a further development or evaluation of the material, and thus should be considered an integral part of the Unit. This is especially true when a problem is referred to in the text. In any case it is wise to look at each problem for its information.

In reading this book it should be remembered that only an introduction to each topic is given, emphasizing the aspects where physics is important. Entire books could be and have been written on the subjects of each Unit. An effort was made to keep the discussion of each topic short enough that the Units can reasonably be assigned as supplementary reading in a standard physics course. Some references are given at the end of each Unit to help those who are interested in further information. These reference lists are not exhaustive, but rather meant to get one started. More extensive bibliographies can be found in the Resource Letters on the environment prepared for the American Association of Physics Teachers: R. H. Romer, "Resource Letter ERPEE-1 on Energy: Resources, Production, and Environmental Effects" Amer. J. Phys. 40, 805 (1972), and John I. Shonle, "Resource Letter PE-1 on Physics and the Environment" Amer. J. Phys. 42, 267 (1974).

In any endeavor such as this the author is indebted to many people. First of all there are the authors of the many texts and articles from which I have learned about the environment. These are too numerous to list. However, any mistakes must be attributed to me alone. I would appreciate hearing about needed corrections in the text. Secondly I must thank the students who bore with me while parts of this book were tried on them. Finally, there are the reviewers of the manuscript: Clyde Zaidins, C. R. Cothern, and David Lazarus.

Suggested Uses of the Units

| Unit | General Area Needed | Texts, see below for key S - Z | A - F | H - R |
|---|---|---|---|---|
| 1 | One dimensional kinematics | 4 | 5 | 3 |
| 2 | One dimensional kinematics | 4 | 5 | 3 |
| 3 | Newton's laws, momentum | 8 | 7 | 8 |
| 4 | Newton's laws | 5 | 7 | 5 |
| 5 | Work, energy, and power | 7 | 8 | 6 |
| 6 | Torque, statics | 3 | 4 | 12 |
| 7 | Rotational dynamics | 9 | 11 | 11 |
| 8 | Gravity | 5 | 15 | 14 |
| 9 | Fluids | 12 | 8 | 15 |
| 10 | Sound | 23 | 23 | 17 |
| 11 | Sound | 23 | 23 | 17 |
| 12 | Kinetic theory of gases | 20 | 13 | 20 |
| 13 | Gas laws and adiabatic expansion | 19 | 13 | 20 |
| 14 | Second law of thermodynamics | 19 | 13 | 21 |
| 15 | Second law of thermodynamics | 19 | 13 | 21 |
| 16 | Electrostatics | 26 | 19 | 25 |
| 17 | Circuits and electric power | 28 | 21 | 27 |
| 18 | R-C circuits | 29 | 21 | 28 |
| 19 | Magnetic fields | 30 | 17 | 29 |
| 20 | Magnetic fields and the law of induction | 33 | 20 | 31 |
| 21 | Fluid flow and transformers | 35 | 21 | 31 |
| 22 | Blackbody radiation and light | 37 | 25 | 39 |
| 23 | Blackbody radiation | 17 | 25 | 39 |
| 24 | Atoms | 43 | 25 | 40 |
| 25 | Heisenberg uncertainty principle | - | 30 | 40 |
| 26 | Nuclear physics | 44 | 22 | - |
| 27 | Nuclear physics | 44 | 22 | - |
| 28 | Nuclear physics | 44 | 22 | - |

The specific texts are:

S - Z: F. W. Sears and M. W. Zemansky, University Physics, 4th ed. (Addison-Wesley, 1970)

A - F: M. Alonso and E. J. Finn, Physics (Addison-Wesley, 1970)

H - R: D. Halliday and R. Resnick, Fundamentals of Physics (J. Wiley and Sons, 1970)

The Appendices can either be considered general information, or the first three could be assigned as regular units as the instructor sees fit.

If this book is used for an environmental course with a physics pre-requisite, the following order of usage within each topic is suggested. The order in which the different environmental topics are taken up is a matter of individual preference, except that it is recommended that transportation follow immediately after air pollution, since the majority of the air pollutants come from transportation. Since this text focuses on the interaction of physics with the environment and on quantitative results, it is suggested that a more general text, such as J. Priest, Problems of Our Physical Environment (Addison-Wesley, 1973), L. Hodges, Environmental Pollution (Holt, Rinehart, and Winston, 1973), or R. H. Wagner, Environment and Man (W. W. Norton, 1971), be used in conjunction.

Air Pollution: Units 24, 4, 12, 16, 13, 23
Transportation: Units 5, 7, 17, 3, 6, 1, 2
Noise and Communication: Units 10, 11, 25
Solid Waste: Unit 8
Energy: Appendices 1, 3, Units 14, 15, 9, 19, 21, 22, 26, 18, 27, 20, 28

The specific texts are:

—— F. W. Sears and M. W. Zemansky, University Physics, 4th ed. (Addison-Wesley, 1970)

—— M. Alonso and E. J. Finn, Physics (Addison-Wesley, 1970)

—— [illegible] and R. Resnick, Fundamentals of Physics (Wiley, 1970)

The appendices contain the consolidated general information, and the [illegible] should be regarded as required [illegible] instructor only.

If only [illegible] a possible [illegible] the following order [illegible] within each topic is indicated. The order [illegible] the environmental [illegible] and [illegible] transportation [illegible] after air pollution since the majority of the air pollutants [illegible] transportation. [illegible]

[illegible] pollution [illegible] environment [illegible]

[illegible] 1973 [illegible] environmental pollution [illegible]

[illegible] and [illegible] W. [illegible] in the [illegible]

[illegible] Air Pollution [illegible]

Transportation [illegible]

Noise and Communications [illegible]

[illegible] 21, 23 [illegible]

# Unit 1
# Transportation I: Traffic Movement

## INTRODUCTION

One of the problems facing cities today is the transportation of people and goods. The main way this is done in most US cities is by motor vehicles. The problems of traffic congestion are too well known to the inhabitants of any large city to need documentation here. In addition, motor vehicles account for the majority of urban air pollution, use up valuable natural resources (especially oil), and cause much death and injury. In this Unit we mostly confine our attention to the problems of traffic flow.

## STOPPING DISTANCES

We need to know the relation between speed and stopping distance for some of the applications which follow. To a reasonably good approximation we can consider a stopping car to be a case of uniformly accelerated motion. This assumes that the brakes are applied steadily, and more importantly, that the brakes continue to function well throughout a single stop. While this latter condition should definitely be the case in all cars for safety's sake, unfortunately it is not always true at higher speeds. This is an area in which federal performance standards have been slow in coming.

With the reasonable assumption of uniformly accelerated motion, our basic equations (in the usual notation) are:

$$x = v_o t + at^2/2 \tag{1}$$

$$v^2 = v_o^2 + 2ax \tag{2}$$

The total stopping distance, which we shall call $x_{ts}$, is composed of two parts: the distance $x_r$ covered during the time that the driver is reacting to the stimulus, and the distance $x_b$ covered while the brakes are applied. During the time that the driver is reacting, the velocity remains essentially constant. We thus use Eqn. 1 with $a = 0$:

$$x_r = v_o t_r$$

where $t_r$ is the reaction time, which typically is around 0.5 sec but may range from 0.3 sec to perhaps 1 sec. The braking distance is most easily found from Eqn. 2, where the final velocity $v = 0$:

$$0 = v_o^2 + 2ax_b$$

$$x_b = -\frac{v_o^2}{2a}$$

One should keep in mind that $a < 0$ for stopping.

The total stopping distance is the sum of the above:

$$x_{ts} = x_r + x_b = v_o t_r - \frac{v_o^2}{2a} \tag{3}$$

Accelerations are often expressed in terms of the acceleration due to gravity $g$. The values of $a$ vary considerably depending on the car, and the tire, brake, and road conditions. Some cars under good conditions may achieve $a = -.9g$. A value of $-.6g$ might be more typical. With wet or icy pavement the values will be much less in magnitude. The following Table shows total stopping distances as a function of initial speeds for various values of $a$ and $t_r$.

Table of Total Stopping Distances

| $a$ | $t_r$ | 20 mph | 40 mph | 60 mph | 80 mph |
|---|---|---|---|---|---|
| -.75g | .5sec | 33 ft | 101 ft | 205 ft | 345 ft |
| -.5g | .3 | 36 | 125 | 268 | 465 |
| -.5g | .5 | 41 | 137 | 286 | 489 |
| -.5g | 1.0 | 56 | 166 | 330 | 548 |
| -.2g | .5 | 82 | 327 | 693 | foolhardy |

We can imagine two possible situations when we consider the safe driving distance between two cars. The first is when it is conceivable that the lead car might stop essentially immediately such as in the case of a head-on collision or a collision with a fixed object. In this case, the spacing between cars should be the total stopping distance $x_{ts}$. This spacing is always the safest, but is seldom actually practiced on crowded roads. The minimum spacing which should ever occur would correspond to the distance covered during the reaction time, $x_r$. The only justification for such a small spacing would be when there was complete assurance that the lead car is moving at the same speed and also takes the entire braking distance before being brought to rest. One does observe in practice distances between cars of even less than $x_r$ -- a very dangerous situation.

## TRAFFIC LIGHT TIMING

Properly synchronized traffic lights which are "green" for sufficiently long times can help the flow of traffic and indirectly reduce air pollution. Some aspects of this are considered in Problem 7.

For traffic to be controlled properly it is necessary for the yellow light to be on for a sufficient length of time that a car can either stop or completely clear the intersection. If a person is farther from the "stop" line than the minimum total stopping distance $x_{ts}$ when the light turns yellow, he can stop in time. If he is less than that distance he should proceed through the intersection at constant speed.

The light must remain yellow long enough for the car to cover $x_{ts}$ and the width of the intersection $w$. Since time is distance divided by speed,

$$t_{yellow} = \frac{x_{ts} + w}{v_o} = \frac{v_o t_r - v_o^2/2a + w}{v_o}$$

$$= t_r - \frac{v_o}{2a} + \frac{w}{v_o}$$

If we assume $t_r = 0.5$ sec and $a = -.5g$, and as an example let $w = 100$ ft and $v_o = 30$ mph $= 44$ ft/sec, we obtain

$$t_{yellow} = .5 - \frac{44}{2(-16)} + \frac{100}{44} = 4.1 \text{ sec}$$

It should be mentioned that if one is moving at less than the "design speed," or the braking acceleration changes because of ice, one can get in a situation in which it is neither possible to stop nor to make it through the intersection during the yellow light. See Problems 3 and 4.

Compare the results of the above model with observations of actual traffic signals for various intersection widths and design speeds.

## TRAFFIC FLOW RATE

New freeways are built and almost as soon as they are opened they are clogged with cars. How many cars can a lane of traffic carry per hour? What is the speed for maximum traffic volume? Can traffic congestion be solved by more freeways? We examine some of the factors involved next.

If the average spacing between cars is $d$ and the average speed is $v_o$, the interval in time between two cars, $T$, is

$$T = \frac{d}{v_o}$$

The number of cars per unit time, $N$, is then

$$N = \frac{1}{T} = \frac{v_o}{d}$$

Various assumptions about the way that the spacing $d$ varies with speed can be made. Here we assume that it is the total stopping distance $x_{ts}$ plus the length of the car. Alternatives are examined in Problems 5 and 6. Note that the length of the car must be included since $x_{ts}$ refers to the distance needed between the back of the leading car and the front of the following car; whereas, $d$ refers to the spacing between corresponding points on the cars. With our assumption the number of cars per unit time is

$$N = \frac{v_o}{x_{ts} + \ell} = \frac{v_o}{v_o t_r - v_o^2/2a + \ell} \qquad (4)$$

$$= \frac{-2a}{-2at_r + v_o - 2a\ell/v_o}$$

where $\ell$ is the length of a vehicle. We rearranged the expression for N for ease of finding the speed which gives the maximum value for N. While we could find $\frac{dN}{dv_o}$ and set it equal to zero (which is the usual procedure in extremum problems), it is simpler if we note that N will have a maximum when the denominator has a minimum. Thus, we find

$$\frac{d}{dv_o}\left(-2at_r + v_o - \frac{2a\ell}{v_o}\right) = 1 + \frac{2a\ell}{v_o^2}$$

If we set the derivative equal to zero, we obtain

$$v_{omax} = \sqrt{-2a\ell}$$

(Don't let the minus upset you -- remember $a < 0$.) The subscript "max" was added to show that this is the speed for maximizing N. Note that $v_{omax}$ is independent of the reaction time in this model. Suppose that $a = -.6g$ and $\ell = 20$ ft. Then

$$v_{omax} = \sqrt{-2(-.6)(32)(20)} = 27.7 \text{ ft/sec} = 19 \text{ mph}$$

Various traffic checks have shown that the maximum flow often occurs for about this speed. Any urban freeway driver knows that when traffic is heavy, the speed drops considerably. We obtain the maximum flow rate by putting the expression for $v_{omax}$ into Eqn. 4:

$$N_{max} = \frac{\sqrt{-2a\ell}}{\sqrt{-2a\ell}\; t_r + 2\ell} = \frac{1}{t_r + \sqrt{\frac{2\ell}{-a}}} \qquad (5)$$

Note that $N_{max}$ increases as $|a|$ increases and as $t_r$ and $\ell$ decrease. For the values $a = -.6g$, $\ell = 20$ ft, and $t_r = .5$ sec, we obtain

$$N_{max} = .5/\text{sec} = 1800/\text{hr}$$

Again, actual traffic measurements show something like this number, but somewhat higher, corresponding to the fact that the actual spacing is usually less than what was assumed above. (See Problems 5 and 6.)

The low capacity of a freeway lane shows the futility of the "more lanes" approach to congestion. Obviously, only a very small fraction of a city of 500 000 to several million can be accommodated by building another lane. Recall that working hours are such that many people want to drive at the same time. In addition, another lane invites more cars to try to use it. The off-ramps, non-freeway streets, and parking spaces remain the same and are unable to cope with additional traffic.

There are alternative models for traffic flow which are more sophisticated (see Ref. 1). One model considers that the response of the driver in the following car depends on the relative spacing and the relative speeds of the two cars. Conclusions are similar to those derived here.

When traffic is heavily congested, instabilities in the flow occur. This means that a minor slow-down gets amplified until there is a complete stoppage of flow (or even collisions). The main cause of instabilities is a driver with a slow reaction time or one who over-reacts to a slow-down in front of him. Unfortunately, the faulty driver almost never is caught by the consequences of his actions. The instability becomes bad only many cars later. Trying to maintain a constant spacing between cars also leads to instabilities, as do sudden changes in the road. It has been shown that limiting the number of cars per unit time can prevent instabilities and actually increase the volume of traffic at peak times. Good driver training and the quick removal of wrecks and disabled cars by helicopter would also help traffic flow.

## REFERENCES

1. R. Herman and K. Gardels, "Vehicular Traffic Flow," Scientific American, Dec., 1963, p. 35.

2. L. Mumford, "The Highway and the City," in The Environmental Handbook, G. deBell, editor. (Ballantine Books, 1970) p. 182.

3. A. T. Demaree, "Cars and Cities on a Collision Course," in The Environment, edited by the editors of Fortune. (Harper & Row, 1970) p. 89.

## PROBLEMS

1. What is the total stopping distance for 30 mph and 60 mph when it is so icy that $a = -.1g$?

2. One should not "over drive his headlights." This means that the safe speed at night should be such that one can stop within the distance that objects

can be seen by the headlights. Measure this distance for your car and find your maximum safe night speed.

3. Suppose the duration of the yellow light on a traffic signal has been set for a speed of 45 mph, intersection width of 80 ft, and an acceleration of -.5g. Can one always either safely stop or clear the intersection when ice reduces $a$ to -.3g, allowing for a reduced driving speed?

4. Consider the problem of "getting caught" by the yellow light if you are going less than the design speed. As speed decreases the total stopping distance decreases but the time to clear the intersection increases. Under what conditions can one be sure of either stopping or clearing the intersection when travelling slower than the speed for which the duration of the yellow light was designed? How much slower can you go? Assume the same $t_r$ and $a$. It is helpful to plot $t_{yellow}$ as a function of $v_o$.

5. What is the maximum traffic flow rate and the speed at which it occurs if one assumes the spacing $d$ to be the length of a vehicle plus the average of the reaction-time distance $t_r$ and the total stopping distance $x_{ts}$? (This assumption more nearly corresponds to the spacing that actually occurs.)

6. Repeat Prob. 5, but let the spacing be the length of a vehicle plus the reaction-time distance. (This is not a safe way to drive.)

7. Since frequent stops and starts both increase air pollution and slow down traffic flow, it is important to have the green light on sufficiently long. Calculate the length of time that a light should be green for $N$ cars to go. Assume that all cars are at rest initially. You will need to consider such factors as the length of the cars, the distance between cars when stopped, the acceleration of cars (which can be assumed constant for low speeds), the reaction time of the drivers, the increased spacing between cars as speed increases, and possibly the speed limit. You may need to make some observations in the field to obtain some of the numbers you need.

# Unit 2
# Transportation II: Some Considerations for Mass Transportation

## BACKGROUND

There are several reasons why mass transportation is now in the limelight. First of all, there are congested streets and freeways, and no place to park once one gets there. We saw in the previous Unit that it is fundamentally futile to try to solve congestion by adding more freeway lanes, since the carrying capacity per lane was only about 2000 cars per hour. Buses can transport more people per hour over a given lane, since they can carry more passengers per vehicle. In any event, additional freeways constitute visual pollution, and there is a limit to parking spaces. Most of the urban air pollution comes from cars. One bus carrying 35 people pollutes much less than 35 cars each carrying one person. Another important consideration is energy consumption. A standard size car with two people in it may achieve about 16 to 28 passenger-miles/gallon. The concept of passenger-miles/gallon is a measure of the overall energy efficiency for moving people. The higher this number, the greater the transportation of people for a given amount of fuel. The passenger-miles/gallon figure is the product of the miles/gallon of the vehicle and the number of people carried. A bus may be able to do 5 times better than the car just mentioned if it is reasonably occupied. Economic factors are also becoming important as the costs of owning and operating a car rise.

In spite of the good reasons for mass transportation, it is well known that most people in most US cities shun it. Some of the reasons are quite valid; others are more imaginary or ones of habit. The main reason to drive your own car is that it goes where you want it to go, when you want it to (barring traffic jams). Buses and trains are often inconvenient in terms of both routing and scheduling. If there are enough buses or trains and enough different lines for really satisfactory convenience, then the system is too expensive, and the savings in fuel and air pollution are largely negated. Mass transportation is usually slow, for reasons which will be explored later. Mass transportation systems are often crowded, dirty, noisy, and uncomfortable as well. Many people prefer automobiles because they are "private." Finally, the automobile has become part of the "American way of life."

What we will do in this Unit is show that it is indeed very difficult to design a mass transportation system to be fast, convenient, and inexpensive. Part of the problem lies with the relatively low density of population per unit area in most cities. The cities in which mass transportation works reasonably well now are the large population centers. Probably the only way mass transportation will succeed elsewhere is by subsidizing it to make it economically attractive, or by making the usage of private automobiles less attractive, perhaps by taxation or even outright prohibition.

AVERAGE SPEED

We will first show that it is not possible for a single system to be both fast and convenient in terms of the spacing of stops. We do this by finding the average speed as a function of the appropriate factors.

Average speed is distance divided by time:

$$\bar{v} = \frac{d}{T}$$

We will suppose that the stops are all spaced the same, and call that distance $d$.

The total time $T$ is made up of the time stopped, $t_s$, the time while accelerating, $t_+$, the time travelling at maximum speed, $t_{max}$, and the time for decelerating, $t_-$. We will assume uniformly accelerated motion for the calculation of $t_+$ and $t_-$. This assumption is rather good for the braking period, but not so good for the acceleration. It will suffice for this model, however. The appropriate general equation is

$$v = v_o + at$$

For the positive acceleration period we have

$$v_{max} = a_+ t_+ \tag{1}$$

and for the deceleration period

$$0 = v_{max} - |a_-|t_- \tag{2}$$

where we use the absolute value of the acceleration for convenience. Separate values for $a_+$ and $a_-$ are appropriate since most vehicles can stop better than they can start. From Eqns. 1 and 2 we have

$$t_+ = \frac{v_{max}}{a_+} \tag{3}$$

and

$$t_- = \frac{v_{max}}{|a_-|} \tag{4}$$

We will drop the absolute value bars around $a_-$ in what follows and just remember we are dealing with the magnitude.

The time at maximum speed is a little more involved. We first of all suppose that $d$ is large enough that maximum speed is reached. See Prob. 1 for the contrary. It is necessary to find the distance covered during the acceleration periods. The relation which is easiest to use is

$$v^2 = v_o{}^2 + 2\ a\ x$$

which leads to

$$x_{\pm} = \frac{v_{max}^2}{2a_{\pm}}$$

The time at maximum speed is the distance covered at maximum speed, which is $(d - x_+ - x_-)$, divided by the maximum speed:

$$t_{max} = \frac{d - x_t - x_-}{v_{max}}$$

$$= (d - \frac{v_{max}^2}{2a_+} - \frac{v_{max}^2}{2a_-})/v_{max}$$

$$= \frac{d}{v_{max}} - \frac{v_{max}}{2a_+} - \frac{v_{max}}{2a_-} \qquad (5)$$

The total time $T$ is thus

$$T = t_s + t_+ + t_{max} + t_-$$

$$= t_s + \frac{v_{max}}{a_+} + \frac{d}{v_{max}} - \frac{v_{max}}{2a_+} - \frac{v_{max}}{2a_-} + \frac{v_{max}}{a_-}$$

$$= t_s + \frac{d}{v_{max}} + \frac{v_{max}}{2} (\frac{1}{a_+} + \frac{1}{a_-})$$

We thus obtain for the average speed

$$\bar{v} = \frac{d}{T}$$

$$= \frac{d}{t_s + \frac{d}{v_{max}} + \frac{v_{max}}{2}\left(\frac{1}{a_+} + \frac{1}{a_-}\right)} \tag{6}$$

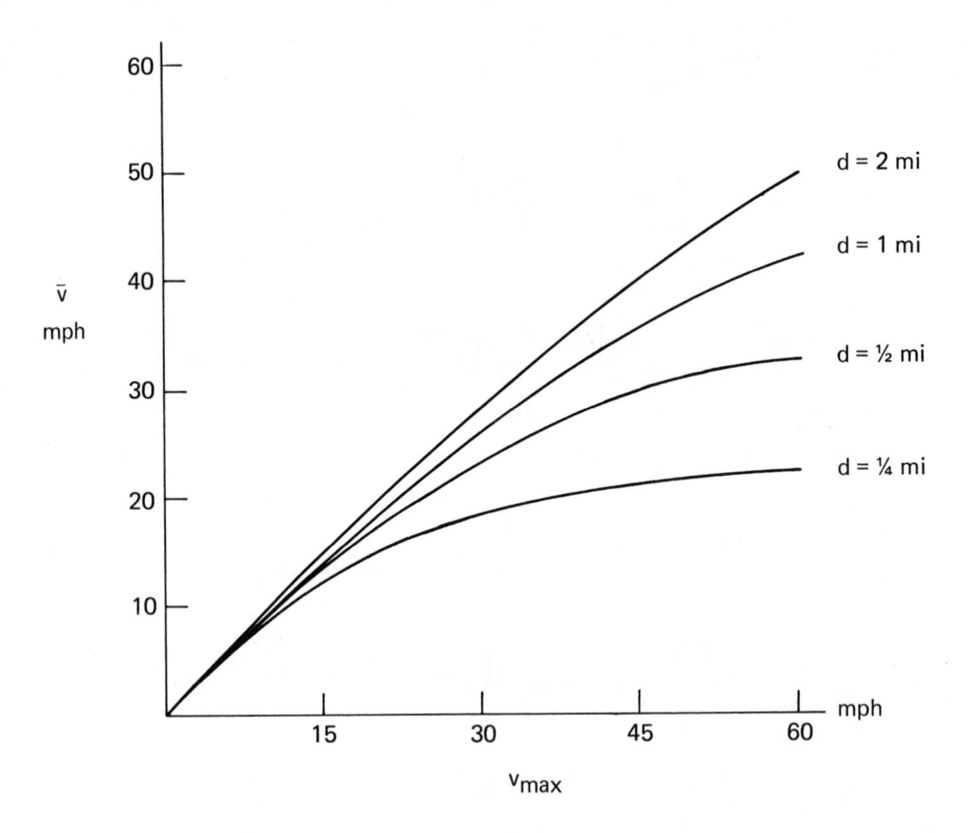


Fig. 1 The average speed as a function of maximum speed and spacing of the stops. The duration of a stop is 10 sec here.

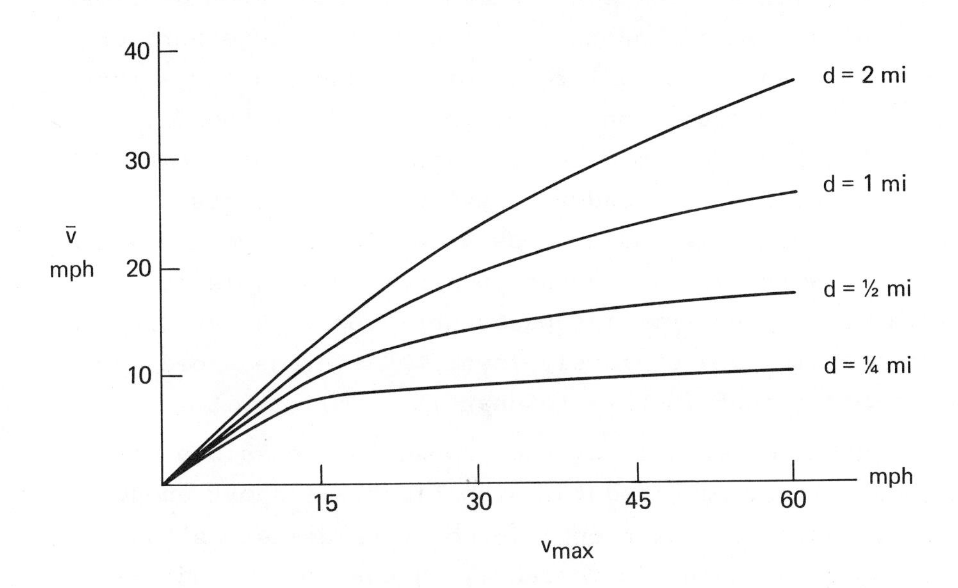


Fig. 2 Same as Fig. 1 except $t_s = 60$ sec.

This equation is sufficiently complicated that it needs discussion and numerical evaluation. In the limit that d is very large, the term $d/v_{max}$ in the denominator makes the other terms negligible and $\bar{v} \to v_{max}$. The spacing d must be about 2 miles or more for this to happen at higher speeds and longer stop times. Further understanding of Eqn. 6 is best obtained by graphing it. We will assume that $a_+ = 4$ ft/sec$^2$ and $a_- = 10$ ft/sec$^2$. Figure 1 shows the results for some selected distances and the assumption that the stopped time is 10 sec, which is appropriate when only a few passengers get on or off. Figure 2 shows the results when $t_s = 60$ sec, more appropriate for times of heavy usage. The evaluation of Eqn. 6 for other cases is left as Prob. 2.

It is clear from the results that high average times are not obtained when the stops are close together, or when the stop time is large. Since stopping for a traffic light is equivalent to stopping for passengers, we see that freedom from traffic lights (as in the case of a train or special roadway), synchronized lights, or bus-controlled lights are essential for a decent average speed. Worse yet is having to stop for signals at separate places from the passenger stops. Heavy traffic and traffic jams obviously lower the average speed more, but this effect is hard to quantify.

Quite respectable average speeds are possible with $d = 2$ miles, provided the stop times are short enough. Unfortunately, most people do not consider stops that far apart as being convenient. On the other hand, as we saw in Unit 1, heavy automobile traffic is not rapid either; the maximum volume was obtained when speeds were about 20 mph. If more traffic tries to use the road, instabilities arise leading to start-stop driving and very low average speeds.

A partial remedy to the problem of not being able to provide both speed and convenience with one system is to have two systems: one high speed with stops perhaps every two miles, and the other a local system feeding into the first. See Fig. 3. One of the chief disadvantages of this system is that most passengers will have to transfer one or more times to reach his destination. Sometimes very long indirect trips are needed to connect two points. For example, consider points $a$ and $b$ in Fig. 3. Even though there is a high speed system, the effective average speed will be very low in such a case. Trains are well suited for the high speed portion since many people can be carried in a single train, and

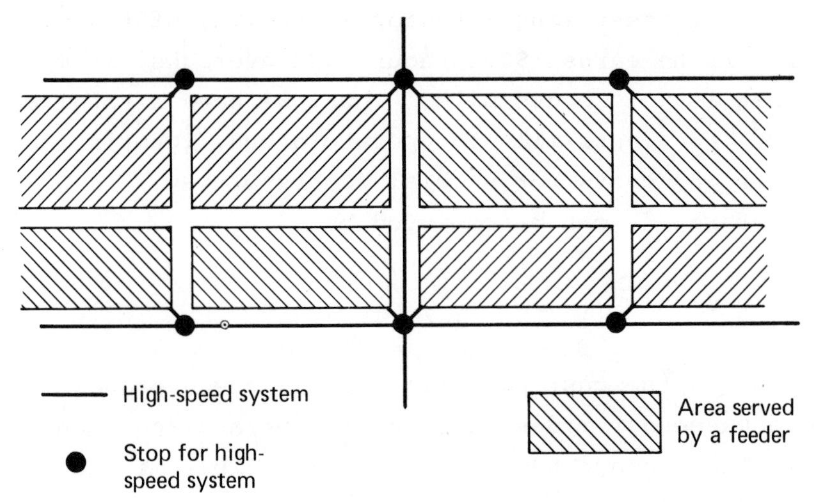


Fig. 3 A combined high speed and feeder system.

fixed-routes are satisfactory. The feeder system would probably use buses, either operating on fixed routes or being dispatched on demand with the aid of a central computer. The latter is sometimes called random-routing. See the references, particularly Ref. 1, for further discussion.

## ESTIMATION OF COST

By having enough buses on enough different routes one could achieve convenient mass transportation. The difficulty lies in the cost. We estimate here the approximate cost of running a bus. A medium sized bus costs about $35,000. If we suppose a 400 000 mile "lifetime" for the bus, we obtain 9 cents/mile for capital investment. If the bus was financed by an 8% bond, interest costs add another 2 cents/mile. Operating and maintenance costs together are about 20 cents/

mile. The largest single factor is the pay of the driver. If he earns \$5.50/hour and averages 15 mph, this cost is 37 cents/mile. Thus, we have a total cost of about 68 cents/mile to operate a bus, ignoring any profit or any costs for special rights-of-way.

The fare $f$ can be computed by

$$f = c\,d/n$$

where $c$ is the cost per mile, $d$ is the average distance a passenger rides, and $n$ is the average occupancy. The average number of passengers per mile needed for a given fare $f$ is thus

$$n/d = c/f$$

With our estimated cost of 68 cents/mile, and a fare of 30 cents, one needs a bit more than two passengers per mile boarding the bus on the average. If the bus stops are located every 1/4 mile, then there must be on the average a passenger at every other stop, day and night, every single bus. The more often the buses run, the harder it is to get the needed patronage. In areas of high population density, where traffic congestion makes the automobile unattractive, it is not hard to achieve averages like the foregoing. But most US cities have had their major growth during the era of the automobile and hence are very spread out. Greater Los Angeles is a prime example. With the sprawl comes low density and thus difficulty in having enough usage of the transportation system to keep the fare low without subsidy.

A similar analysis could be made for trains and subways. Here there are the added factors of right-of-way acquisition or tunnel building, and track cost. These forms of transportation require much more capital outlay. On the other hand, labor costs are less per passenger, since many more people can ride in one train. Clearly, very high usage is needed for trains or subways to be economical. We will not treat the cost quantitatively here.

## DISCUSSION

While many people think in terms of new technologies, such as monorails, air-cushion buses, and linear induction trains, the main problems of mass transportation still are in planning for convenience and the economics. The choice of method is secondary, and follows after questions of routing, population density along the route, and cost have been answered. One can divide the methods into two categories: buses over public roads, or transportation over exclusive rights-of-way. The latter would include tired vehicles on special roads, trains, subways, and monorails. Because of the high costs involved with exclusive rights-of-way, they are feasible only when the population is great enough to insure adequate usage. Fortunately trains, etc., are not needed except in this case. Buses can be of various sizes from the minibus up to large articulated ones. They can have either fixed routes or on-demand flexible routes, as mentioned before.

One important factor to keep in mind when designing a mass transportation system is that not everyone lives in the suburbs and works or shops downtown. Many systems are designed as if this were the case. We might call it the "upper-middle-class syndrome" in design. To be sure it is much easier to design and build a system that will get people from a, b, c, and d to H than it is to get people from e to G or f to I.

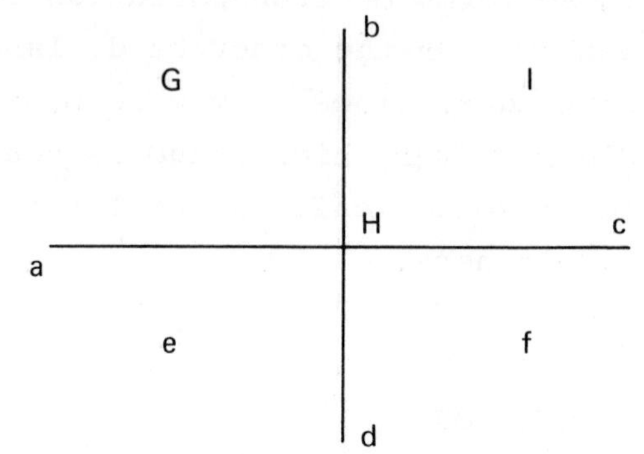


Another thing to keep in mind is that the demand for transportation peaks around 8-9 am and again 5-6 pm. The system must be able to handle the peak load, yet be economical the rest of the time. The equipment and personnel to run it largely are idle during non-rush hours, creating a large overhead expense. Thus, the capacity must be set according to the peak demand, and the economics according to the average demand.

Perhaps one of the best compromises among some of the conflicting requirements for mass transportation is to use buses driving on exclusive reserved lanes of existing public roads. This way costs can be kept reasonable, while average speeds are not lowered by traffic jams. This system has usually worked well where tried. One could easily add devices whereby the buses could trigger traffic signals to be always green for them.

## REFERENCES

1. T. R. Stone, Beyond the Automobile, (Prentice-Hall, 1971).

2. A. T. Damaree, "Cars and Cities on a Collision Course," in The Environment, edited by the editors of Fortune. (Harper & Row, 1970).

3. W. F. Hamilton, II, and D. K. Nance, "Systems Analysis of Urban Transportation," Scientific American, July, 1969.

4. J. Priest, Problems of Our Physical Environment, (Addison-Wesley, 1973) ch. 8.

## PROBLEMS

1. Derive a new expression for the average speed assuming that $d$ is sufficiently short that maximum speed is not reached.

2. Evaluate Eqn. 6 numerically for some other cases. For example: how much would higher accelerations help? How much worse is it if the stops are 0.1 mile apart? Be careful to use compatible units. It is suggested that ft, ft/sec, and ft/sec$^2$ be used.

3. Equation 6 appears to imply that as $v_{max}$ gets large enough the average speed will decrease

because of the term $\frac{v_{max}}{2}\left(\frac{1}{a_+}+\frac{1}{a_-}\right)$ in the denominator. This clearly violates common sense. Resolve this apparent inconsistency. Hint: examine the assumptions made.

4. Try to design a mass transportation system in which no one has to walk more than three blocks, which uses high speed and feeder systems, which minimizes the number of transfers, and yet is economically feasible. One appreciates the problems of mass transportation much more after trying actual designs.

# Unit 3
# Transportation III: Traffic Safety

## PERSPECTIVES

Usually, the first things that come to mind when environmental problems are mentioned are air or water pollution or biocides. Automobile accidents are seldom mentioned. Yet the automobile death rate far exceeds the probable deaths from pollution. The London smog disaster of December, 1952, where perhaps 4000 died or the Donora, Penna., incident of 1948 where 20 died are often cited as evidence of the dangers of air pollution. It has been estimated that perhaps a few hundred deaths a year can now be attributed to air pollution in the US. In contrast, the automobile death rate is more than 55 000 per year in the US alone. More than 2 000 000 are injured a year. When seen in this perspective, we must indeed count traffic safety as an important social, if not environmental, goal.

PRELIMINARY CONSIDERATIONS

To get a feel for the energy involved in a car crash, let us calculate the equivalence of the kinetic energy of a car driving 60 mph in terms of dynamite. We will take 4000 lb as the weight of the car. Now

$$E_k = mv^2/2$$

So we must convert weight to mass and express the speed in appropriate units. If we use English units,

$$m = W/g = 4000/32 = 125 \text{ slugs}$$

and

$$60 \text{ mi/hr} = 88 \text{ ft/sec}$$

Thus,

$$E_k = (125)(88)^2/2 = 4.84\text{x}10^5 \text{ ft-lb}$$

The heat of combustion for dynamite is 1.29 kcal/gm. The conversion between ft-lb and kcal is

$$1 \text{ ft-lb} = 3.24\text{x}10^{-4} \text{ kcal}$$

So the kinetic energy of a 60 mph car is

$$E_k = 4.84\text{x}10^5 \text{ ft-lb } (3.24\text{x}10^{-4} \frac{\text{kcal}}{\text{ft-lb}})$$

$$= 1.57\text{x}10^2 \text{ kcal}$$

Thus, the kinetic energy is equivalent to

$$\frac{1.57\text{x}10^2 \text{ kcal}}{1.29 \text{ kcal/gm}} = 121 \text{ gm}$$

of dynamite.

When a car crashes into a fixed barrier this kinetic energy is suddenly converted into other forms. It is like having 121 gm (about 1/4 lb) of dynamite exploding in the front of your car.

The point about the relation between a head-on crash between two similar cars going at the same speed and the crash of a single car into a barrier causes much confusion. These two cases are essentially equivalent if both are inelastic (no rebound). There are two ways to see that this is so. The first one notes that two cars at a given speed have twice the kinetic energy as a single one, but there are two cars to dissipate this energy by crumpling up. The more satisfactory argument notes that the crash of two cars would be about the same if a piece of sheet steel is inserted between them at the point of impact. Since this piece of metal is not moved during the impact (for similar cars at the same speed), the crash is the same as if the piece of metal were fastened to a fixed object.

The case of a head-on crash of dissimilar cars or unequal speeds is now considered, since many people are turning to smaller cars to get better gasoline mileage. We will again suppose an inelastic collision in which the two cars remain together. Although cars will often have their brakes applied before and during a crash, this external force is sufficiently smaller than the forces associated with the crash that to a good approximation linear momentum is conserved. Let the masses be $m_1$ and $m_2$, and the velocities beforehand be $v_1$ and $-v_2$ and the velocity after $v_f$. Then from momentum conservation we have

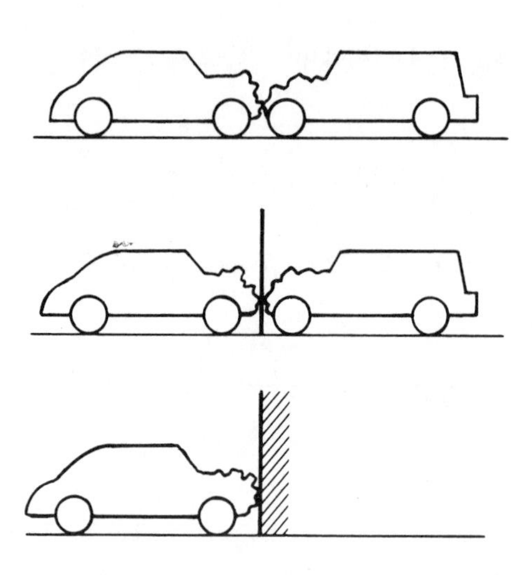

Fig. 1 The collision of a car with a fixed barrier is the same as the head-on collision of two similar cars.

$$m_1 v_2 - m_2 v_2 = (m_1 + m_2) v_f$$

This equation may be solved for $v_f$ and the changes in velocity may be found:

$$\Delta v_1 = v_f - v_1 = - m_2 (v_1 + v_2)/(m_1 + m_2)$$

$$\Delta v_2 = v_f - (-v_2) = m_1 (v_1 + v_2)/(m_1 + m_2)$$

Since the time interval is the same for both, the average accelerations are proportional to the above velocity changes. We first consider the equal mass case. The magnitudes of the changes in velocity are then equal to the average speed, $(v_1 + v_2)/2$. So the higher speed car has a smaller acceleration, and the lower speed car, a larger acceleration, than if each separately hit a fixed barrier. In the case of unequal masses, the smaller car has the larger acceleration. As an example, consider a Cadillac (4,840 lb) and a VW (1,808 lb). The magnitudes of the velocity changes are

$$|\Delta v_{Cad}| = .27 (v_1 + v_2)$$

$$|\Delta v_{VW}| = .73 (v_1 + v_2)$$

Thus, the VW has over 2.6 times the acceleration of the Cadillac. Accident statistics definitely show a higher rate for serious injuries or death in small cars, but these data are subject to a variety of interpretations. One is to eliminate small cars. Another is to eliminate big cars. Others include the need for better design on small cars and requiring large cars to drive slower.

In some collisions the cars separate afterwards, so the collision is partially elastic. A fully elastic head-on collision of equal cars would send the cars rebounding backwards with interchanged speeds. Thus, they would have even greater accelerations than in the inelastic case. So it is important to design energy absorbing material rather than springy material into a car. The cases of glancing and broadside collisions, sometimes with a second impact, are too varied for general discussion.

## SURVIVING A CRASH

We assume in this section that car accidents will occur and inquire into methods for surviving them. Since most accidents are caused by human error and it is unlikely that this can be programmed out of people, the above assumption is reasonable. We examine some of the ways in which the occurrence of accidents can be reduced in a following section.

It is difficult to obtain specifications for the maximum magnitude of acceleration the human body can stand without major injury. This is an area in which experimentation with live volunteers must be limited. Cadavers do not respond the same as live people do. It is virtually impossible to deduce the accelerations after-the-fact in actual crashes. It is known from volunteers that people can survive about 20 g's acceleration with no significant injuries. The various federal standards and proposed standards for crashes imply from 10 g's to 80 g's. A higher value of acceleration can be tolerated if it is for a short period of time, so a weighted average acceleration times the duration of the impact, called the severity index, has been proposed. The acceleration is expressed in terms of $g$:

$$a = -\alpha g$$

and the following integral is defined:

$$\text{Severity index} = \int_{\substack{\text{duration}\\ \text{of crash}}} \alpha^{5/2}\, dt$$

Because of the 5/2-power weighting given to $\alpha$, high values of $\alpha$ can occur for only very short intervals of time. A somewhat arbitrarily proposed specification limits the severity index to 1000 for the head in a crash from 30 mph. To get a feel for the magnitude of acceleration involved, let us assume that $\alpha$ is constant. Then

$$\text{Severity index} = \alpha^{5/2}\, T$$

where $T$ is the duration of the crash. We obtain $T$ from

$$v = v_o + at$$

which becomes

$$0 = v_o - \alpha g T$$

thus,

$$T = v_o/\alpha g$$

and

$$\text{Severity index} = \alpha^{5/2} \frac{v_o}{\alpha g} = \alpha^{3/2} \frac{v_o}{g}$$

For a value of 1000, and for $v_o = 30$ mph = 44 ft/sec, we obtain

$$1000 = \alpha^{3/2} \frac{44}{32}$$

or

$$\alpha = \left(\frac{32(1000)}{44}\right)^{2/3} = 81$$

Considering the range of proposed safe accelerations, we adopt here the conservative value of $a = -30g$ for a good chance of surviving. Note that this value implies a force of 30(150) = 4500 lbs acting on a 150 lb person!

If we suppose that a person is properly constrained (more on that later), then the value $a = -30g$ must apply to the passenger-carrying portion of the car. We now calculate the stopping distances for various speeds that correspond to the chosen acceleration. We start from

$$v^2 = v_o^2 + 2ax$$

and obtain

$$0 = v_o^2 + 2(-30g)x$$

or

$$x = v_o^2/(60g) = v_o^2/1920 \qquad (1)$$

if feet and seconds are used. On evaluating Eqn. 1, we obtain

| $v_o$ | x |
|---|---|
| 30 mph | 1.01 ft |
| 60 mph | 4.04 ft |
| 90 mph | 9.09 ft |

These figures correspond to the minimum distance that the front portion of the car must yield in order to keep the acceleration of the passengers at a tolerable level. For the above values to be valid, the accelerations must be constant. From

$$F = ma$$

we see that constant acceleration implies constant force. This is difficult to obtain with conventional design. A

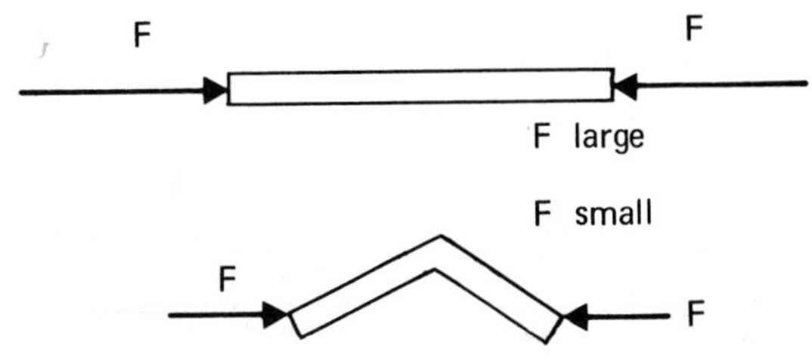


Fig. 2 The variation of force as a member crumbles.

straight structural member can take a very large force. Eventually, it will start to buckle and then the force needed for further deformation is very small, resulting in a very nonuniform force. By proper design a controlled yielding of crumpling of the front structure to provide approximately a constant force can be achieved. Care must be taken to keep things like motors, wheels, steering columns, etc., from intruding into the passenger compartment. Another way of achieving a constant force is by having the front portion of the car connected by hydraulic pistons to the remainder.

It is rather easy to obtain a controlled crumpling distance of 1 foot, suitable for 30 mph. The 4 feet required at 60 mph would be much more difficult to achieve, but probably obtainable. Any value larger than this would probably lead to cars too bulky for city driving.

The above safe controlled crumpling distances which apply to the passenger compartment will do no good if the passenger is not constrained so that he shares this motion. Without constraint a passenger will continue moving forward with essentially undiminished speed. During the time he is flying through the air, the car may have come to rest so that he will hit a stopped windshield (or some other part of the car) while still travelling at the original speed, thereby encountering enormous accelerations. See Prob. 3 for a consideration of this point. Thus, some form of constraint is needed as well as a controlled yield design for full safety.

There are presently two methods of constraint: seat belts and air bags. The former is well proven, the latter still under development. The main reason for constraint is to insure that the passenger shares the deceleration rate of the center part of the car. Seat belts do this admirably. A further reason for constraint is to keep a person inside the car. It has been estimated that one's chances of surviving are only about 10% as great if one is ejected from a car. In spite of this, many people argue against wearing seat belts because one _sometimes_ survives a crash by being thrown from the car. Fire is also sometimes given as a reason not to wear seat belts, on the assumption that it would take longer to get out of the car. In the first place, fire is quite rare in a crash (about 0.2%). In the second place, it would be better to be conscious and uninjured with seat belts than mangled without them in case of a fire. Seat belts hold a driver in place so that he possibly can control a car after a glancing collision or skid.

A detailed study in Sweden showed no deaths below 60 mph when seat belts were in use, but fatalities occurred at as low as 12 mph without them. In general, injuries were reduced by about 50% through the use of seat belts. A study in Utah showed similar results. The use of seat belts reduced the chances of death or serious injury by 50%, and more minor injuries were reduced 30% to 40%. Victoria, Australia, now requires seat belts to be used. There was a drop in fatalities by 13% in rural areas and 24% in urban areas, in spite of the fact that only about 3/4 of the population were complying with the law. Yet only about 1/4 of Americans use seat belts regularly. Many use them only on long trips or at high speeds. However, most accidents occur near home, and seat belts are most effective at preventing death at lower speeds. While statistics are very incomplete, the use of shoulder straps with seat belts may increase the chances of survival another 60%. Shoulder straps should never be used by themselves.

The main argument for air bags is that they require no action on the part of the occupants of a car. They are for people who do not want to take the trouble to protect themselves. Air bags inflate in about 0.040 sec to 0.050 sec after the start of a crash. Problem 2 examines how far an unconstrained person can move during this time. In some cases, the bag does not inflate fast enough, since passengers are often already thrown forwards because of panic braking, a skid, or a glancing collision which occurs before the main collision which triggers the air bag. Air bags are designed for frontal collisions. They offer little or no protection for broadside collisions or roll-overs. These latter are more dangerous than the frontal collisions. There is a real reliability problem: will they work when needed after perhaps several years of non-use, and will they not go off when not needed? One consultant firm estimated that the noise associated with the rapid inflation of air bags might cause permanent hearing damage to 25% of the people exposed to the noise. Another difficulty comes from the fact that an air bag offers more and more resistance as a person flys into it farther. A nonuniform acceleration results, with a high final value. The person may also be pushed violently back into the seat back, with further injury.

## MORE ON STOPPING

People are seldom taught how to stop properly in an emergency. Usually, drivers react to crises by jamming on their brakes, locking the wheels. This is especially easy to do with power brakes. Brakes should never be applied so hard as to lock the wheels (which means that the wheels are no longer turning but sliding over the pavement). There are several reasons for this. In the first place, the coefficient of sliding friction is less than the coefficient of static friction. When the wheels are still rolling, the bottom of the tire is at rest, and the static case prevails. This means that a greater frictional force is possible and thus a shorter stopping distance than when the tires are sliding. (One should keep in mind that the frictional force of a tire is a very complicated subject and only crudely approximated by the usual simplified "laws" of friction.) The question of "weight transfer" to the front tires while braking is very important and will be considered in Unit 6. Because of this weight transfer, the rear wheels often lock up first. If the car is slightly deviated from a straight line motion while the rear wheels are locked up, a skid usually results. (The reason for this is left to the reader.) When a car skids, it often leaves the lane it has been travelling in, which on a crowded road can be very bad. If on the other hand the front wheels have been locked and are sliding, it is impossible to steer the car. The tire skids on straight independently of the direction it is pointed. Putting these factors together we see that locked wheels increase the chances of an accident by increasing the stopping distance, promoting skids, and causing a loss of directional control. One should apply the brakes just as hard as will not lock the wheels. Then it is still possible to control the car and possibly steer clear of an accident. One often hears advice to "pump" the brakes on ice. The idea behind this is that the brakes get released periodically, stopping any lock up. However, a series of locked and released brakes is not as effective as a steady non-locked braking.

Unfortunately, it is very difficult to know how to apply the optimum braking, especially with power brakes. A prudent driver will practice hard braking some safe place where an unintended skid will cause no difficulties. Airplanes have had anti-locking devices for years. Such devices could be installed on all cars and would certainly prevent many minor accidents (especially those in heavy traffic). At present the cost is rather high -- around $300 to $400 for effective four-wheel sensing and control.

Problem 5 discusses the often bad effects of applying brakes while cornering.

## OTHER FACTORS AFFECTING TRAFFIC SAFETY

We will now briefly mention some other important factors in safety. First is drunken driving. It is now estimated that alcohol is involved in over 50% of all fatal accidents. An unknown portion of car fatalities are actually cases of homocide and suicide, but it has been estimated to be a non-negligible fraction. Good driver licensing and training would help. Driving tests should include high speed driving, emergency stopping, and response tests on an accident simulator. Student drivers must be taught defensive driving and what to do in the case of an emergency. Skid practice should be mandatory. Car handling quality is another area in which improvements are needed. Many cars steer sluggishly, react poorly to sudden turns, and have poor brakes. Automatic avoidance devices are technically possible. They might operate by a sonic radar and a sensor for telling lateral position on the road. If the relative speed is too great for the spacing, speed would automatically be decreased. Devices can tell when a driver is sleepy, drunk, or otherwise responding too slowly. Unfortunately, most effective control devices are expensive.

REFERENCES

1. Patrick M. Miller, "The Crashworthiness of Automobiles," Scientific American, 228 No. 2, 78 (Feb. 1973)

2. J. Rose (editor), Technological Injury, (Gordon and Breach)

3. Road & Track

   Many issues have relevant articles. See particularly Jan., 1968, Dec., 1969, Feb., 1972, July, 1972, Feb., 1973.

PROBLEMS

1. Falling from what height is equivalent to crashing at

   a) 30 mph b) 60 mph?

2. Assume that 0.040 sec is needed for an air bag to be significantly inflated, and that the passenger compartment of the car accelerates with -30g. Calculate how far relative to the passenger compartment an unrestrained passenger will have moved during this time for various initial speeds.

3. If there is no constraint of any kind a passenger will continue going until he hits the windshield. Under what conditions will the windshield already be at rest when he hits it? Assume a constant acceleration. Find a general expression and evaluate it numerically for several cases.

4. Calculate how stopping distances change when going up or down a hill. Find a general expression, and evaluate it numerically for several typical cases.

5. One should never apply the brakes while rounding a corner fast. The reason is that the tires may be near the limit of their adhesion in just providing the centripetal force. The additional force needed for reducing the speed may make the total force exceed the maximum frictional force, thus causing a skid. Work out the details of such a possibility, assuming that $\mu = .7$ adequately characterizes

the frictional force of the tires. Use a general radius $R$ and speed $v$, and find what braking acceleration can safely be used in various cases. Neglect weight transfer.

# Unit 4
# Air Pollution I:
# Mechanical Removal of Particulates

## INTRODUCTION

One traditionally thinks of air pollution in terms of chimneys belching smoke. Most of the earlier efforts and legislation concerned visible air pollution, which is largely due to particulate matter. Unfortunately, the air pollution which is most damaging to people, plants, and structures is in the form of invisible gases: the nitrogen oxides, the sulfur oxides, ozone, and the many products of photochemical reactions.

However, smoke and particulates do have their undesirable effects. The most obvious is their soiling effect. It has been estimated that "dirty" air costs the nation an extra $1 billion per year in cleaning bills. Good visibility is important to aviation. Particles in the air can serve as condensation nuclei for rain drops and fog. There is some statistical evidence to indicate local increases in rain near large sources of particulates.

Particulates can cause lung irritation. Synergistic effects have been noted between particles and $SO_2$. The word "synergism" means that the effect of two things together is greater than the sum of the effects of each alone. Particles, especially carbon, can adsorb large quantities of gases on their surfaces and thus carry them deep into the lungs. This can be especially bad in the case of carcinogens. Atmospheric particles are also known to aggravate bronchitis.

Particles in the atmosphere might cause large scale climate changes. The details are not well understood, and there is still debate as to the direction of the change. One school of thought holds that particles will reflect or scatter sunlight back into space, thus leading to a cooling effect. One estimate indicated a possible decrease in global temperature of 3.5° C if the aerosol concentration increases by a factor of four from present levels. See Reference 2. (The term "aerosol" means a solid particle or liquid drop which remains in the air for a long time.) Another line of reasoning points to the absorption of radiation by the particles leading to a heating effect. Indeed, particles injected into the stratosphere by the Mt. Agung eruption increased the stratospheric temperature about 6° C, but there was no measurable change in the ground-level temperature. The absorption characteristics of the particles for infrared radiation are important and not well known now. It is possible that particles could contribute to the "greenhouse effect." (See Unit 23.)

We should note that Nature is also putting particles into the air by volcanoes, dust storms, and ocean spray at a greater rate than man is. It has been claimed that just three volcanic eruptions (Krakatau, 1883, Mt. Katmai, 1912, and Hekla, 1947) have put more particles into the air than all of man's activities.

The main man-made sources of particulates and the number of tons of each placed in the atmosphere during 1968 are: burning of fossil fuels, $8 \times 10^6$ tons; industry, $7 \times 10^6$ tons; agricultural burning, $2 \times 10^6$ tons; and transportation and refuse disposal each with $1 \times 10^6$ tons. Particulates emitted by automobiles are at first too small to be visible but coagulate to become visible.

This Unit considers mechanical methods of removing particulate matter. Thermal-gradient precipitators are discussed in Unit 12 and electrostatic precipitators in Unit 16.

## GRAVITATIONAL SETTLING

The main ways in which particles are removed from the atmosphere are by gravitational settling and wash-out by rain and snow. Gravitational settling can also be used as a control method for larger particles. We examine here the rate of fall of small particles. The hypothesis of free fall is <u>not</u> valid for small particles. In other words, we cannot neglect air resistance. Over a certain size range the drag force is proportional to the speed of the particle and its size. If we assume a spherical shape and ignore a small correction term, then Stoke's law is valid:

$$F_{drag} = - 6 \pi \eta r v$$

where $\eta$ is the viscosity of air, $r$ is the radius of the particle, and $v$ is its speed. We can ignore the small buoyancy force due to the air. Newton's second law is then

$$F_{net} = mg - 6\pi\eta rv = ma \tag{1}$$

If the particle starts from rest, then initially $a = g$. As the particle speed increases, the acceleration decreases. Eventually, the particle will be going fast enough that the drag force will equal the gravitational force in magnitude and the acceleration will be zero. This speed is called the terminal velocity and can be found by setting $a = 0$ in Eqn. 1:

$$mg - 6\pi\eta rv_{term} = 0$$

or

$$v_{term} = \frac{mg}{6\pi\eta r} \tag{2}$$

We can relate the mass to the particle size and density $\rho$ by

$$m = \frac{4}{3}\pi r^3\rho$$

If we put this value into Eqn. 2 we obtain

$$v_{term} = 2g\rho r^2/(9\eta)$$

We will evaluate this expression in cgs units, since $\eta$ is usually expressed in terms of Poise, which is a cgs unit. For air at 20° C, $\eta = 1.81 \times 10^{-4}$ Poise. We will take $\rho = 1$ gm/cm$^3$ (water) and $\rho = 8$ gm/cm$^3$ (iron). The results are shown in Fig. 1. We rather arbitrarily

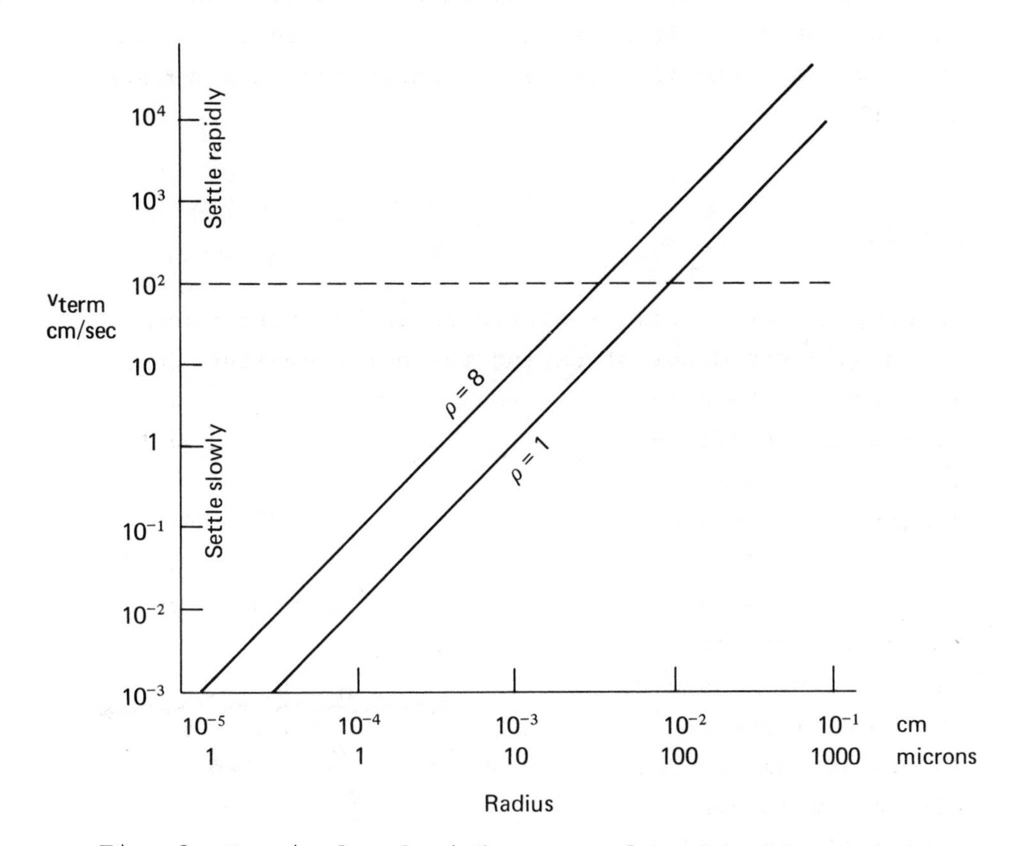


Fig. 1 Terminal velocities as a function of particle radius.

adopt $v_{term}$ = 100 cm/sec = 1 m/sec as the dividing line between rapid and slow settling.

Particle sizes smaller than those shown in Fig. 1 do not settle at all, since air cannot be considered a continuous fluid on that scale, and molecular bombardment must be considered. This gives rise to Brownian motion.

Particles larger than about $10^{-2}$ cm can be controlled by ducting the air into a large settling box, provided the speed of the air flow is not too great. A set of baffles helps the settling by directing the particles downward. In most cases the particle sizes are too small and the flow rates too large for this method to work.

FILTERS

An obvious way to remove particles is by filtration. The simple expedient of having the holes smaller than the size of the particles is not effective, since such a filter gets clogged too easily and provides too large an impediment to the flow of air. Instead, multiple fibers or layers are used. The air molecules can easily flow around the fibers, but the

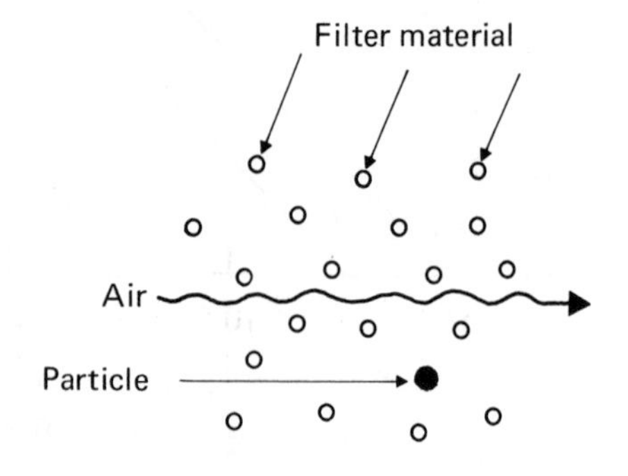


Fig. 2 Action of a filter.

particles with their much larger mass per particle tend to go in straight lines by the principle of inertia and are intercepted by the filter material. Even with a relatively open structure a significant amount of power is often needed to force the air through the filter. See Problem 3. Clogging of the filter remains a problem. Sometimes mechanical agitation is used, and sometimes a reverse blast of air to unclog the filter. Relatively little of the collected material goes back into the air stream since it has coagulated into much larger particles. Filtration is successful for particles from about $10^{-5}$ cm to about $10^{-2}$ cm. There is a lack of filter material which can stand the high temperatures found in some stacks.

## CENTRIFUGAL FILTERS

Inertia can be used to remove larger particles ($10^{-1}$ cm to 10 cm). This is commonly done by the cyclone separator shown in Fig. 3. The particles tend to go in a straight line and hit the outside wall; whereas, the air can be easily deflected due to the much smaller mass of the air molecules. The air passes out the center and particles settle out the bottom by gravity.

## WET SCRUBBERS

Particles may be removed by passing the dirty air through a fine spray of water. The particles are intercepted and removed by the water drops. The diagram (Fig. 4) is purely schematic; many different configurations are used. Wet scrubbers are effective for $10^{-2}$ cm to $10^{-5}$ cm. An additional advantage of this method is that it is partially effective for removing some of the gaseous pollutants as well. The sulfur oxides are readily soluble in water and the nitrogen oxides are to some extent. One real difficulty with wet scrubbers is that one then has the problem of disposing of polluted water. Another difficulty is that the exhaust gases are cool and humid and may quickly settle to the ground rather than dispersing.

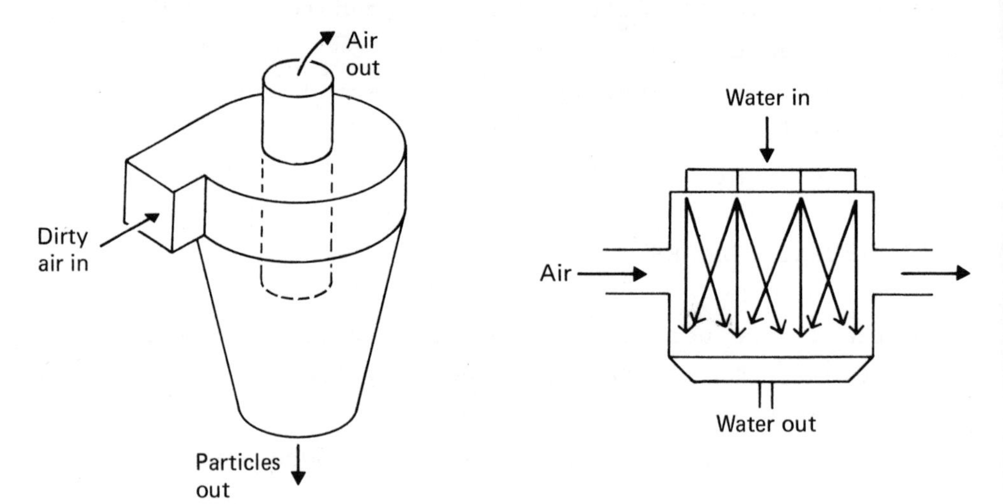


Fig. 3 Centrifugal filter.

Fig. 4 Schematic of a wet scrubber.

REFERENCES

1. A. Stern (editor), Air Pollution, (Academic Press, 1968) Vol. III.

2. S. I. Rasool and S. H. Schneider, "Atmospheric Carbon Dioxide and Aerosols: Effects of Large

Increases on Global Climate," Science, 173, 138 (1971)

3. Man's Impact on the Global Environment, (MIT Press, 1970).

PROBLEMS

1. Verify Fig. 1.

2. Suppose that there is only a 5% chance for a particle to be filtered out by a given layer of filter material. How many layers are needed to remove 99% of the particles?

3. Calculate the power needed to filter $10^5$ $ft^3$/min of dirty air. Assume that the pressure drop through the filter is 10 to 25 lb/$ft^2$. Does your answer depend on the area of the filter (assuming a fixed pressure drop, which is not too realistic)?

4. Use Fig. 1 to explain why rain falls but fog does not.

# Unit 5
# Transportation IV: Automotive Energy and Power Requirements

## INTRODUCTION

The automobile is the source of the majority of the urban air pollution, with few exceptions. Estimates vary between about 50% and 90%, on a weight basis. Almost all of the carbon monoxide in the air comes from automotive sources. Many cities sometimes have CO levels high enough to be a health hazard. Unburned hydrocarbons and nitrogen oxides are both plentiful in automotive exhaust, and these are the main raw ingredients for the eye-irritating photochemical smog. There are presently about 90 000 000 cars and almost 20 000 000 trucks and buses in the US. These figures are rising about 3% per year. Something must be done. There are three main routes by which the air pollution problem from automobiles may be reduced:

1) Reduce the emissions from internal combustion engines.

2) Substitute mass transportation for the private automobile.

3) Use another power source for automobiles.

The first alternative is not likely to be satisfactory for more than perhaps 10 years, since the number of cars is rising. The second alternative has its difficulties, as discussed in Unit 2. We thus consider the third one. In this Unit we estimate the power and energy required to operate a car which must be satisfied by any alternative power source.

## FRICTIONAL DRAG

Friction is ever-present, so some power is needed even for steady speed on level ground. This frictional drag arises primarily from air resistance, rolling friction, and lubricated sliding friction. Little is contributed by the dry sliding friction usually discussed in basic text books. The main friction present is velocity dependent. We thus must seek the form of this dependence. We will use as our basic data the speed *vs* time curve while coasting for a Checker cab (*Road & Track*, *Road Test Annual*, 1969). This car was chosen as it represents a typical car size. Furthermore, the coasting curve was very similar to those of many other cars. The curve is reproduced in Fig. 1. We immediately note that it is *not* a straight line, thus ruling out a constant acceleration, consistent with the above discussion. We can calculate average accelerations for each 5 second interval from

$$\bar{a} = \frac{v_2 - v_1}{5}$$

The results are shown plotted against the average velocity for each time interval in Fig. 2. We note that there is some irregularity in the points of Fig. 2; however, the data are reasonably well fit by a straight line. Thus, we can write

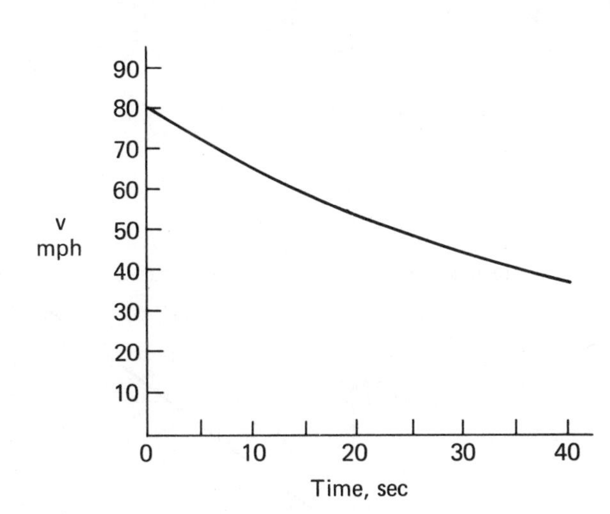


Fig. 1 Speed vs time while coasting for the Checker (from Road & Track magazine, by permission).

$$a \propto -v$$

or

$$a = -kv$$

Realistically there is probably a constant term and a small quadratic term also present. The data are adequately represented by Eqn. 1 for our purposes, however. We can rewrite Eqn. 1 as

$$\frac{dv}{dt} = -kv \qquad (2)$$

which means that the derivative of velocity is proportional to the velocity. The only function which has its derivative proportional to the function is the

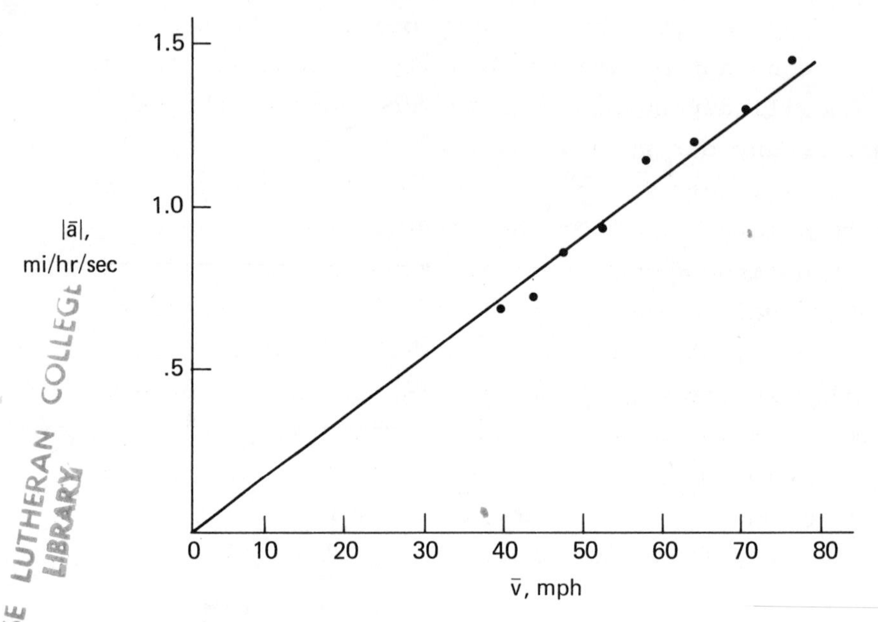




Fig. 2 Acceleration as a function of speed for the Checker.

exponential function. (This may be proven by integrating Eqn. 2 by separation of variables or from the general theory of linear differential equations.) We thus assert that

$$v = v_o e^{-kt} \tag{3}$$

The reader can show that Eqn. 3 is consistent with Eqn. 2 by differentiating Eqn. 3. Anytime that one variable is an exponential function of another, a semi-logarithmic plot of the former against the latter will yield a straight line. This is left as an exercise for the reader (Problem 5).

Before we can calculate numerical data on the power and energy requirements of automobiles we must evaluate $k$, the constant of proportionality in Eqn. 1. The value of $k$ will depend on the size and shape of the car. However, any car with a coasting curve the same as Fig. 1 will have the same value of $k$. Rather surprisingly a large range of the cars in the 1969 Road Test Annual do indeed have similar curves. Some fall off more gradually, indicating lower drag and a smaller value of $k$, while others have velocity decreasing more rapidly, indicating a larger value of $k$. From the 'best' straight-line fit of the points in Fig. 2, we obtain a value of $k = 0.018\ \text{sec}^{-1}$. Note that this value is independent of the units used for measuring speed.

The drag force is obtained from Newton's second law:

$$f = ma = m(-kv) = -\,mkv \tag{4}$$

where the minus sign indicates the force is directed oppositely to $v$.

## POWER REQUIREMENTS

We now calculate the power required from

$$P = Fv$$

We consider the general case of a car accelerating on a hill with drag. Various special cases such as constant speed on the level can then be obtained from the general case. The important forces acting on a car are shown in Fig. 3, where $F_m$ is the force due to the motor, and $f$ is the drag force. From Newton's second law we obtain

$$F_m - |f| - mg \sin \theta = ma$$

or, using Eqn. 4,

$$F_m = ma + mg \sin \theta + mkv$$

$$= m(a + g \sin \theta + kv)$$

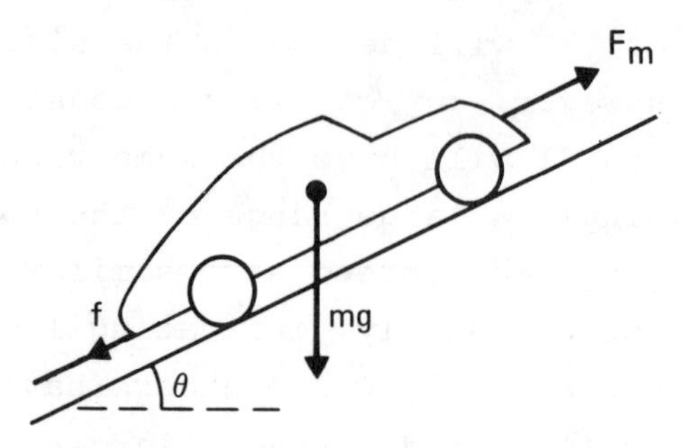

Fig. 3 Forces on a car.

If we use English units, $g = 32 \text{ ft/sec}^2$ and with the representative value of $k = .018 \text{ sec}^{-1}$, we obtain

$$F_m = m(a + 32 \sin \theta + .018v)$$

The power needed is thus

$$P = F_m v = mv(a + 32 \sin \theta + .018v) \tag{5}$$

This equation refers to the power needed at the driving wheels, which is generally only 60% - 80% of the advertised engine brake horsepower.

For constant speed on level ground Eqn. 5 reduces to

$$P = .018mv^2$$

Let us use the weight of the Checker cab, 3840 lb. Its mass is then

$$m = 3840/32 = 120 \text{ slugs}$$

At 60 mph = 88 ft/sec, the car needs

$$P = .018(120)(88)^2 = 16700 \text{ ft-lb/sec}$$

Since 1 hp = 550 ft-lb/sec, we have

$$P = 16700/550 = 30.4 \text{ hp}$$

The same car at 60 mph on a 5% upgrade needs 61 hp and on a 10% upgrade needs 92 hp. A car must be able to accelerate at a reasonable rate if it is to pass another car safely. If we take the acceleration to be 4 $ft/sec^2$, then at 60 mph we need 107 hp. The student is invited to verify these figures and find the power in some other cases. (Problem 4). Note that since power is not recoverable in an internal combustion car, values of $P$ less than zero are not meaningful.

We thus see that a reasonable case can be made for 100 hp at the wheels, or an engine rating of perhaps 150 hp to allow for losses and the fact that the peak power is obtained at a single engine speed only. Horsepower ratings in excess of this do not make much sense environmentally.

The previous remarks apply to a car intended for the open road. There is a growing awareness that the "open road" car is not desirable for city driving. It is too large, consumes too much gas, takes up too much parking space, and pollutes too much. Seldom are there more than two people in a car in town. The maximum speeds on city streets are seldom over 40 mph (freeways excepted, of course). Long distances are not involved. Thus, the idea of a smaller "town car" for urban use has arisen. Let us see what the power requirements of such a car might be. We assume a loaded weight of 2080 lbs (mass = 65 slugs), the same drag coefficient, a maximum acceleration of 6 ft/sec$^2$ on the level, and a 10% hill climbing ability. While a town car would be smaller than a Checker, it would probably not be aerodynamically shaped, so the same drag coefficient is reasonable. The power required for climbing a 10% hill at a steady 40 mph (= 58.7 ft/sec) is

$$P = mv(a + 32 \sin \theta + .018V)$$

$$= 65\ (58.7)(0 + 32 \sin 5.7° + .018(58.7))$$

$$= 16180 \text{ ft-lb/sec} = 29.4 \text{ hp}$$

The power required for accelerating at 6 ft/sec$^2$ on the level at 30 mph (= 44 ft/sec) is

$$P = 65(44)(\ 6 + 0 + .018(44))$$

$$= 19425 \text{ ft-lb/sec} = 35.3 \text{ hp}$$

Sometimes the power required is expressed in terms of specific power, which is defined as power/lb. Sometimes the total vehicle weight is used and sometimes the power plant weight. If we use the former,

$$P/mg = v(a/g + \sin \theta + .018v/g)$$

For constant speed on the level this becomes

$$P/mg = 5.62\times10^{-4} v^2$$

where $P$ is in ft-lb/sec and $v$ in ft/sec. If we express $P$ in hp and $v$ in mph, we have

$$\text{specific power} = 2.20\times10^{-6} v^2 \tag{6a}$$

For $P$ in watts and $v$ in mph, we have

$$\text{specific power} = 1.64\times10^{-3} v^2 \tag{6b}$$

## ENERGY REQUIREMENTS

We can obtain the energy required for a given trip either from

$$E = Pt$$

or

$$E = W = Fd$$

Either approach of course gives the same result. We adopt the former here. For simplicity we first assume a trip of distance $d$ at constant speed $v$ on level ground. The time is thus

$$t = d/v$$

As we have seen in this case

$$P = .018mv^2$$

Thus,

$$E = Pt = .018mv^2(d/v)$$

$$= .018\ mvd$$

This equation shows that the energy consumed for a given distance is proportional to the speed. Since the energy is derived from gasoline, one might suppose that gas mileage would be inversely proportional to speed. At higher speeds this is approximately true. A careful test on one car was made by Road & Track (reported in the December, 1973, issue). The results showed an inverse relation (within about 3% accuracy) between speed and mileage for speeds between 50 mph and 80 mph for steady driving on the level. At 40 mph and slower the mileage was about constant. This deviation from the results of our theory is due to lower engine efficiency at slow speeds and the neglect of a constant term in the drag force.

Again, we can compute on the per-pound basis, or specific energy:

$$E/mg = .018vd/g$$

If the specific energy is in kilocalories/pound of car weight, the speed in mph, and $d$ in miles, we obtain

$$\text{specific energy} = 1.41\text{x}10^{-3}\ vd \tag{7}$$

The unit kcal/lb was chosen since the energy content of fuels is often expressed in terms of kcal. The coefficient in Eqn. 7 becomes $1.64\text{x}10^{-3}$ for W-hr/lb (suitable for electric cars).

The foregoing discussion assumed steady driving speeds. We now consider the energy needed while accelerating. We will assume a constant power level P, even though engine power varies with speed. In this case, we still have

$$E = P\ t$$

Thus, our problem is one of finding the time needed for an acceleration from rest to $v_f$. We cannot use the results of uniformly accelerated motion, since Eqn. 5 shows that acceleration decreases as speed increases for constant power levels. We start from Eqn. 5, which for level roads is

$$P = mva + .018mv^2$$

We solve this for a:

$$a = \frac{dv}{dt} = \frac{P - .018mv^2}{mv}$$

This last equation can be integrated by separation of variables

$$\int_0^{v_f} \frac{mv\,dv}{P - .018mv^2} = \int_0^t dt$$

The integration yields

$$t = \frac{1}{.036} \ln \left( \frac{P}{P - .018mv_f^2} \right) \tag{8}$$

Thus, the energy needed is

$$E = \frac{P}{.036} \ln \left( \frac{P}{P - .018mv_f^2} \right) \tag{9}$$

For small values of $v_f$, Eqn. 9 reduces to the expected result $E = mv_f^2/2$.

As a numerical example, consider the 0 - 60 mph acceleration of a 3200 lb car with 100 hp:

$$E = \frac{55000}{.036} \ln \left( \frac{55000}{55000 - .018(100)88^2} \right)$$

$$= 4.47\text{x}10^5 \text{ ft-lb}$$

The final kinetic energy is

$$E_k = mv^2/2 = (100)88^2/2 = 3.87\text{x}10^5 \text{ ft-lb}$$

Thus, about 15% more energy than that required for the kinetic energy is needed to accelerate the car in this case.

## ENGINE EFFICIENCIES

The typical overall efficiency of an automobile engine is about 22%. The internal combustion engine is first of all a heat engine and as such has basic limits to its efficiency. Actual engines are less efficient than this theoretical limit. Much of the energy is wasted in the form of heat which passes out the exhaust pipe and radiator. Energy is lost in internal friction of all the moving parts, both in the engine and along the drive train. Automatic transmissions tend to have higher losses than manual transmissions due to the fluid coupling. Running an air conditioner will soak up more energy. The complete overall efficiency of a car is considered in Probs. 9 and 10.

## CONSUMPTION OF RESOURCES

Transportation uses about one quarter of the total energy consumed in the US. Over $92 \times 10^9$ gallons of fuel were used by motor vehicles in 1970. This corresponds to over half of the total US production of oil. The best present estimates indicate that we have already probably passed the peak production rate of oil in the US, and that 80% of all the US oil will have been pumped by the year 2000. In the long run the lubricating capability of oil may be much more valuable than the temporary convenient energy content. In addition large quantities of steel go into cars every year. About 200 million tires are produced annually.

Both worn-out tires and cars constitute a severe disposal problem. About one-quarter of the land in an urban area is devoted to automobiles.

## REFERENCES

1. "Control Techniques for Carbon Monoxide, Nitrogen Oxide, and Hydrocarbon Emissions from Mobile Sources," NAPCA, Publication AP-66, (1970, USGPO) ch. 7.

2. Road & Track. Various Road Test Annuals.

## PROBLEMS

1. Measure the drag coefficient for a car yourself. Select an untravelled level area, let the car coast in neutral, and note the speed every few seconds. This will be easier if there are two people. There should be no wind, and a trial should be made going in each direction to average over any slight ups and downs. Do an analysis similar to that given.

2. Derive an expression for the power and energy requirements if the frictional drag has constant and quadratic terms as well as the linear term used here.

3. In order for traffic to merge smoothly onto a freeway from an on-ramp, there must be an acceleration lane long enough to allow the entering cars to

reach the traffic speed. Eqn. 5 shows that the acceleration increases as power increases. Calculate how long acceleration lanes must be for a speed change from 30 mph to 60 mph for 40, 80, and 120 hp cars. Assume a weight of 3000 lbs. Since the acceleration is not constant for a fixed power level, you should approximate the actual case by assuming uniformly accelerated motion with an average value for the acceleration, which may be obtained from Eqn. 5 by using the average speed.

4. Numerically evaluate the power and energy requirements for several additional cases.

5. Eqn. 3 indicates that velocity while coasting varies exponentially with time. If this is the case, a plot of log(v) vs t will be a straight line. Make such a graph from the data in Fig. 1. The easiest way to do this is to use semi-logarithmic paper. Evaluate the constant k from your graph.

6. Find an expression for the top speed of a car on level ground as a function of power from Eqn. 5. Numerically evaluate it for several cases and compare with actual top speeds. Note that the quoted horsepower is usually considerably larger than the delivered horse power at the wheels. Also one should have a drag term proportional to $v^2$ for the best representation of the drag at high speeds. How does the top speed vary with power if there is a quadratic term in the drag expression? High speeds are very costly in terms of power needed.

7. Find the drag coefficient k for other cars from data given in Road & Track. Note that more recent tests have dropped the coasting data.

8. Compare the specific power and specific energy given in Eqns. 6 and 7 with the following graph taken from Ref. 1. Note that in the graph given below the weight refers to the power plant weight, not the total car weight as in Eqns. 6 and 7. In the graph the power plant was assumed to be 500 lbs in a 2000 lb car.

9. Calculate the overall efficiency of a 3840 lb car, if it can obtain a mileage of 15 mi/gal at 60 mph. The energy content of a gallon of gasoline is $3.17 \times 10^4$ kcal. Assume the drag coefficient is .018.

10. Do the same as in Prob. 9, except for a 3000 lb car that can obtain 25 mi/gal at 60 mph. Assume a drag coefficient of .016 in this case.

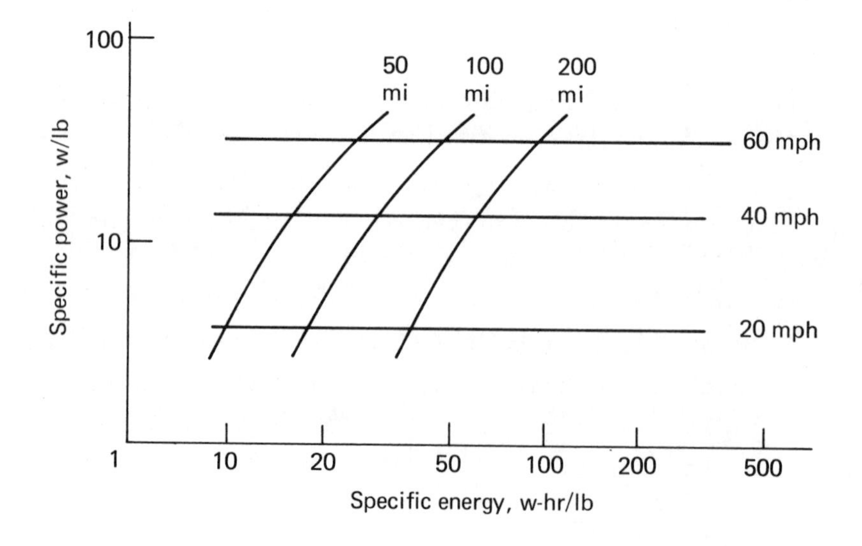

# Unit 6
# Transportation V: Weight Transfer

In Unit 3 we considered the adverse effects to safety of locking up one's brakes, especially when either the front or rear wheels lock up separately. In this Unit we consider some of the factors affecting the lock up of the front or rear brakes. If the braking system is in good condition, the main factors are the weight distribution in the car and the so-called weight transfer effect when the car is accelerating. Since these are both variable factors, it is impossible to achieve optimum proportioning of braking effort between front and rear wheels in all cases. Weight transfer is also an important design factor for cornering and on hills.

Why does a wheel lock up and start sliding instead of rolling? Basically it is a question of torques due to the brakes and friction. When the wheel is not sliding, its angular velocity will be related to the speed of the car by

$$\omega = v/r$$

and the angular acceleration will be related to the linear acceleration by

$$\alpha = a/r$$

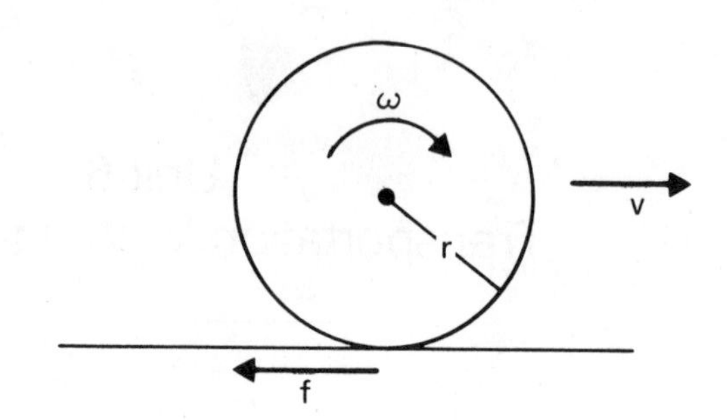


Fig. 1 Sign convention showing positive senses.

When braking, the angular acceleration is provided by the difference between the torque provided by the frictional force $f$ and the torque due to the brakes. We adopt a sign convention such that $v$ and $\omega$ as shown in Fig. 1 are positive. Then $a$ and $\alpha$ are both negative quantities. The net torque is related to the angular acceleration by

$$\tau_f - \tau_b = I\alpha$$

or

$$rf - \tau_b = I\alpha$$

where $I$ is the moment of inertia of the wheel. Rearranging, we obtain

$$\tau_b = rf - I\alpha$$

A static frictional force is a variable quantity with a maximum which can be approximately related to the normal force by

$$f \leq \mu N$$

We thus have

$$\tau_b \leq \mu r N - I\alpha \qquad (1)$$

as the condition for the wheel to roll without slipping. Recall that $\alpha$ is negative in this expression. The angular acceleration can be expressed in terms of $a$ and thus in terms of the total frictional force; see Problem 1. If $\tau_b$ exceeds the above value, the magnitude of the angular acceleration will increase above that required by $\alpha = a/r$, and the wheel will quickly cease rotating. The important point here is that as the normal force on a given wheel decreases, it is easier for the brakes to lock up a wheel.

When a car accelerates or decelerates, the normal forces on the wheels change. This is what is meant by the term "weight transfer."

We now calculate how the normal forces change when the brakes are applied. Figure 2 defines the quantities used. Directions to the right will be considered positive for this treatment. We apply Newton's second law

$$f_1 + f_2 = ma = \frac{W}{g} a \qquad (2)$$

Since the car is not accelerating vertically, the sum of the vertical forces is zero:

$$N_1 + N_2 - W = 0 \qquad (3)$$

The car as a whole is not rotating, so the sum of the torques on the car must be zero. Since the car is accelerating, we must compute the torques about the center of mass of the car. The weight of the car thus produces no torque, but the forces $f_1$, $f_2$, $N_1$, and $N_2$ do. Torque is force times the perpendicular distance from the origin to the line of the force. If we consider counter-clockwise torques as positive, we obtain

$$f_1h + f_2h + N_2(b - d) - N_1d = 0 \qquad (4)$$

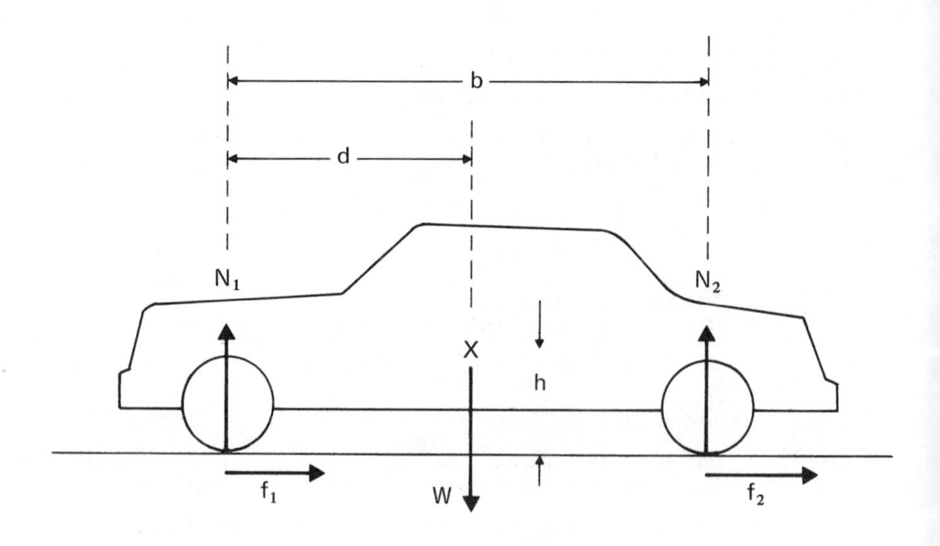


Fig. 2 Forces on a car. The cross represents the center of gravity of the car. $f_1$ and $N_1$ refer to both front wheels together, and similarly for the back wheels.

We solve Eqns. 2, 3, and 4 for $N_1$ and $N_2$ to obtain

$$N_1 = \frac{W(b - d + ha/g)}{b} \tag{5}$$

$$N_2 = \frac{W(d - ha/g)}{b} \tag{6}$$

We are interested in the proportioning of the maximum brake effort between the front and rear brakes. It can be shown that the $I\alpha$ term in Eqn. 1 is negligible compared to the $\mu rN$ term (see Prob. 6). Thus, the torques due to the brakes should be proportional to the respective normal forces if the tires are the same size (see Prob. 7). We obtain

$$\frac{\tau_{b1}}{\tau_{b2}} = \frac{N_1}{N_2} = \frac{(b - d) + ha/g}{d - ha/g} \tag{7}$$

Recall that in this expression a positive value for $a$ represents braking. Eqn. 7 gives the ideal proportioning of the braking torques. The proportioning of the forces applied to the brake mechanisms will depend on many factors, such as whether drum or disk brakes are used, sizes of the drums or disks, number of pads or shoes, piston sizes, and so on. It is readily evident from Eqn. 7 that the ideal proportioning depends on the rate of acceleration and the location of the center of mass. The higher the acceleration and the farther forward the center of mass, the larger the front wheel braking torque should be. Generally, the center of mass shifts rearward as a car is more heavily loaded. What the manufacturers must do is set the proportioning for a compromise loading and acceleration. Variable proportioning is not yet practical. Thus, a car which is lightly loaded ($d$ smaller) or braking very hard will have too much of the braking effort at the rear and will tend to lock up the rear wheels first. A heavily loaded car ($d$ larger) or one stopping with a small value of $a$, as on ice, will probably lock up the front wheels first. These factors must be taken into account for safe driving.

As an example of the size of the weight transfer let us assume $W = 4000$ lb, $b = 10$ ft, $d = 5$ ft, $h = 2.5$ ft, and $a = .6g$.

$$N_1 = \frac{4000(5 + 2.5(.6))}{10} = 2600 \text{ lb}$$

$$N_2 = \frac{4000(5 - 2.5(.6))}{10} = 1400 \text{ lb}$$

For comparison, $N_1 = N_2 = 2000$ lb when the acceleration is zero. There is thus a 30% change in the normal forces in this example when accelerating.

## REFERENCE

1. _Road & Track_, August, 1971, p. 63.

## PROBLEMS

1. Pursue the analysis of Eqn. 1 by expressing $\alpha$ in terms of the car's acceleration and the radius of

the wheel (valid only before wheel sliding starts). Then relate $a$ to the total frictional force to obtain an expression for $\tau_b$ in terms of $\mu$, $N$, $r$, and $I$. Evaluate the maximum $\tau_b$ using reasonable estimates.

2. Find the weight transfer to the outside wheels when rounding a curve of radius $R$ at a speed $v$.

3. Suppose a car is sliding sideways on a road, and it slides into a low curb. Under what conditions will the car roll over?

4. What is the change in the front and rear normal forces when a car is going uphill or downhill at a steady speed?

5. Why are drag racers always rear-wheel drive?

6. Show that $I\alpha$ is much smaller than $\mu rN$ in Eqn. 1 by using reasonable estimates of the quantities involved.

7. Most cars are fitted with the same size tires front and rear. Sometimes owners change the wheels so that the rear tires are much larger than the front. Discuss the effects of the change on brake performance.

# Unit 7
# Transportation VI: Flywheel Cars

## INTRODUCTION

One way to curb air pollution caused by motor vehicles is to replace the internal combustion engine by some other energy source. One alternative which is receiving some attention now is a rapidly spinning flywheel. If steel is used for the flywheel, weight considerations limit the range of a car to about 20 miles. However, recent developments in fibers and composite materials are bringing the energy storage capabilities into a practical range for automobiles and in-town buses.

Several problems remain to be solved. The main obstacle is that the energy to set the flywheel in motion has to come from somewhere. Most likely, electric motors will be used, so a vast increase (50% - 100%) in the electric generation capacity, with all the attendant environmental problems, would be needed if all cars were to be powered by flywheels. Another problem concerns an effective transmission to couple the flywheel to the driving wheels.

Flywheels could also be utilized in a car as a temporary energy storage device in conjunction with an engine (probably gasoline). At first glance it might seem that there would be no advantage to employing the flywheel if an ordinary internal combustion engine must also be running. The reason that an environmental gain is made is that, while the average power required by a car is low, certain cases such as accelerating or going up a hill require large power outputs for relatively short periods of time. For example, in Unit 5 we saw that only 30 hp was needed for steady 60 mph driving, while accelerating from 60 mph at 4 $ft/sec^2$ required 107 hp. The power system of a car must be able to meet these peak power demands. If there is no energy storage, the engine must be large enough for these relatively rare peak periods. If there is stored energy which can be drawn on to meet the peak demands, the engine need meet only the lower average power demand. A flywheel, even one of steel, can reasonably provide the needed energy storage. Not only is the power required of the engine reduced, but with a proper coupling system, the engine can run more nearly at a steady speed. It is considerably easier to reduce the emissions of a fixed-speed engine than one which must go from idling to sudden high speed power bursts.

## ENERGY STORAGE BY A FLYWHEEL

We start with the expression for the energy of a rotating body:

$$E_k = \frac{1}{2} I\omega^2$$

where $I$ is the moment of inertia and $\omega$ the angular speed in radians/sec. The moment of intertia can be written as

$$I = cMR^2$$

where $M$ is the total mass, $R$ a linear dimension such as length or radius, and $c$ a constant which depends on the particular geometry. For example, $c = 1/2$ for a solid disk. There are obvious limits to the moment of inertia set by total weight and physical size. There is also a limit to $\omega$. Circular motion implies acceleration. As the angular speed increases, the acceleration increases as the square of the angular speed

$$a = \omega^2 r$$

Since an acceleration requires a force, the force increases with the square of the speed. This force in a solid object comes from the internal tensile strength of the material. Eventually there will be an angular speed reached at which the required force will exceed the capabilities of the material, in which case the object breaks.

We will now calculate the maximum angular velocity for a rotating rod. Consider a small portion of the rod lying between $x$ and $x + \Delta x$, with mass $\Delta m$. We suppose this piece to be small enough that it is approximately a point mass. Thus

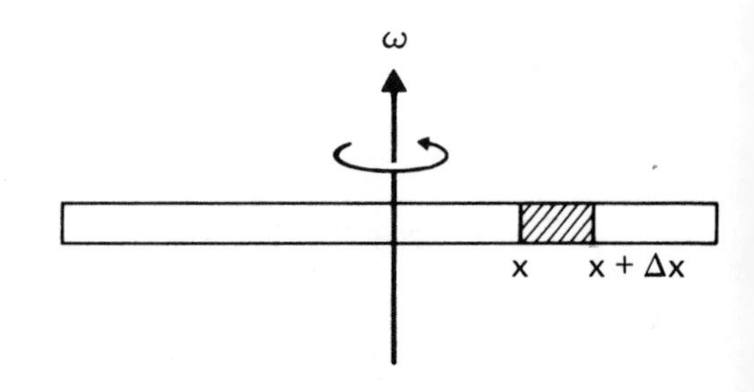

we can apply Newton's second law:

$$F_{net} = (\Delta m)a = -(\Delta m)\omega^2 x \tag{1}$$

The minus sign is needed since the acceleration is directed towards the center, which is in the negative direction. The net force will be the difference between the internal tensions at $x$ and $x + \Delta x$:

$$F_{net} = F(x + \Delta x) - F(x)$$

We can relate the mass $\Delta m$ to the mass density $\rho$ and the cross-sectional area $A$ by

$$\Delta m = \rho A \, \Delta x$$

Putting these last two expressions into Eqn. 1, we obtain

$$F(x + \Delta x) - F(x) = -\rho A \omega^2 x \Delta x$$

or

$$\frac{F(x + \Delta x) - F(x)}{\Delta x} = -\rho A \omega^2 x$$

If we let $\Delta x \rightarrow 0$, the left hand side becomes $dF/dx$, and the equation becomes

$$\frac{dF}{dx} = -\rho A \omega^2 x$$

We now integrate both sides with respect to $x$.

$$\int \frac{dF}{dx} dx = -\int \rho A \omega^2 x dx$$

Which becomes

$$F = -\rho A \omega^2 \int x dx = -\rho A \omega^2 (x^2/2) + C$$

where $C$ is the arbitrary constant of integration. Since the force is zero at the outer end where $x = R$, we may evaluate $C$:

$$0 = -\rho A \omega^2 R^2/2 + C$$

or

$$C = \rho A \omega^2 R^2 / 2$$

Thus

$$F(x) = \rho A \omega^2 (R^2 - x^2)/2$$

The maximum force, as one might expect, is at the center where $x = 0$:

$$F = \rho A \omega^2 R^2 / 2 \qquad (2)$$

A similar result may be obtained for a rotating solid disk. Tensile strengths are expressed in force per unit area. If we let $T$ stand for the tensile strength, and equate $T$ to the force divided by the area in Eqn. 2, we get an expression for the maximum angular velocity:

$$T = \frac{F}{A} = \frac{1}{2} \rho \omega_{max}{}^2 R^2 \qquad (3)$$

A more detailed treatment which considers the elastic behavior of the material before ultimate failure gives results which differ from Eqn. 3 only by a numerical factor near unity. We thus have

$$\omega_{max} = \sqrt{\frac{2T}{\rho R^2}} \qquad (4)$$

The maximum energy stored by a flywheel is thus

$$E_k = \frac{1}{2} I \omega_{max}{}^2$$

$$= \frac{1}{2} (cMR^2)\left(\frac{2T}{\rho R^2}\right)$$

$$= c(M/\rho)T$$

Since

$$M = \rho V$$

where $V$ is the volume, we obtain a surprisingly simple result:

$$E_k = cVT \tag{5}$$

The reader should check that the right hand side of Eqn. 5 does indeed have the dimensions of energy. Note that the maximum energy stored depends only on the volume and the tensile strength, and not directly on the mass of the flywheel. Thus one seeks high strength, low density materials for a flywheel.

Let us rewrite Eqn. 5 in terms of the energy density of the flywheel

$$\text{Energy/weight} = \frac{cVT}{mg} = \frac{cT}{\rho g}$$

where $\rho$ is mass/unit volume, and thus $\rho g$ is weight/unit volume. Some of the new fiber materials have densities around $125\ \text{lb/ft}^3$ and tensile strengths around $5\text{x}10^5\ \text{lb/in.}^2 = 7.2\text{x}10^7\ \text{lb/ft}^2$. So the stored energy density for a disk ( $c = .5$ ) is

$$\begin{aligned} \text{Energy/lb} &= (.5)(7.2\text{x}10^7)/125 \\ &= 2.88\text{x}10^5\ \text{ft-lb/lb} \\ &= 108\ \text{W-hr/lb} \end{aligned} \tag{6}$$

Steel, because of its somewhat lower tensile strength and higher density, can provide only about one-seventh as much energy density. Fused silica can provide about 4 times more than the result in Eqn. 6, but it is difficult to manufacture.

## APPLICATION

We now compare the energy storage capabilities of a flywheel given by Eqn. 5 with the energy requirements of cars given in Unit 5. For steady driving on the level, we had

$$E = .018mvd$$

where m was mass, v the speed, and d the distance. Let v = 60 mph = 88 ft/sec, m = 100 slugs (corresponding to a weight of 3200 lb), and d = 200 miles = $1.056 \times 10^6$ ft. Then

$$E = 1.67 \times 10^8 \text{ ft-lb}$$

From Eqn. 6 we can find the weight of a flywheel which could supply this much energy:

$$\text{Weight} = \text{Energy}/(\text{Energy/lb})$$
$$= 1.67 \times 10^8 / 2.88 \times 10^5 = 581 \text{ lb}$$

This calculation ignores losses in the transmission system. The efficiency should be at least 80% since the form of energy remains the same. If we put the upper limit on weight at 500 lb, and include an 80% efficiency, then $1.15 \times 10^8$ ft-lb are available. This is enough for 138 miles at 60 mph or 200 mi at 41 mph for a 3200 lb car.

It must be realized that it takes longer to set a flywheel spinning with maximum energy than it does to fill up the gas tank on a car. Problems 7 and 8 consider aspects of this problem.

We now turn to the idea of a flywheel as a temporary storage of energy to supplement a lower power internal combustion engine. The energy needed for accelerating from 40 mph to 70 mph for a 3200 lb, 100 hp car can be found from Eqn. 9, Unit 5. The energy to go from $v_1$ to $v_2$ is the difference in energies to go from rest to $v_2$ and from rest to $v_1$:

$$E_{v_1 \to v_2} = E_{v_2} - E_{v_1}$$

$$= \frac{P}{.036}\left[\ln\frac{P}{P-.018mv_2^{\,2}} - \ln\frac{P}{P-.018mv_1^{\,2}}\right]$$

$$= \frac{P}{.036}\ln\frac{P-.018mv_1^{\,2}}{P-.018mv_2^{\,2}}$$

For the figures given above

$$E_{40\to 70} = 4.63\text{x}10^5 \text{ ft-lb}$$

Let us take an efficiency of .8 in the transmission, and allow for enough energy storage for two such accelerations. Then the energy needed is

$$(4.63\text{x}10^5)(2)(1/.8) = 1.15\text{x}10^6 \text{ ft-lb}$$

A flywheel for this much energy would weigh only 4 lb for fiber and 27 lb for steel, which is quite reasonable. The question of energy storage needed for climbing hills is left to Prob. 1. Therefore the use of a flywheel for energy storage could reasonably reduce the maximum needed power output of the engine and thus the pollution.

One severe engineering problem is the method to couple the energy of the flywheel to the car's wheels. The flywheel would be spinning quite rapidly--perhaps 15000 to 50000 rpm. As energy is taken from the flywheel, it naturally slows down, perhaps at a time while the car's speed is increasing. Thus a continuously variable gear ratio is needed. See Prob. 6. One possible coupling would be through an electrical generator, but then the cost and weight get large and the efficiency might be less. Another problem might be with the angular momentum of the flywheel, and a possible "gyroscopic" effect when the car tries to change direction. This is investigated in Prob. 5. One answer to this is to have two counter-rotating flywheels. The rapidly spinning flywheel would have to be well enclosed to prevent lethal fragments from flying out if the flywheel were to break due to material failure or an accident.

## REFERENCES

1. J. McCaull, "A Lift for the Auto" Environment, Dec. 1971, p. 35.

2. R. F. Post and S. F. Post, "Flywheels" Scientific American, Dec. 1973, p. 17.

## PROBLEMS

1. Find the energy needed to climb a hill of length $d$ and angle $\theta$ at a steady speed $v$. Include the drag term. Evaluate the expression for several reasonable cases and find the weight of a flywheel needed to provide this much energy.

2. Find the maximum rotational speed in rpm for a flywheel from Eqn. 4 for fiber and for steel. $T = 2.8\text{x}10^5$ lb/in.$^2$ and $\rho = 480$ lb/ft$^3$ for steel.

3. What weight steel flywheel would be needed to supply the energy for a 100 slug car to go 200 miles at a steady 60 mph? See Prob. 2 for data on steel.

4. One of the main drawbacks of solar power is that the sun may not shine every day. One possible energy storage method would be the use of flywheels. Calculate how large a flywheel would be needed to store 48 hours' worth of energy for a 800 Megawatt power plant.

5. Show that the angular momentum of a flywheel rotating at maximum speed is

$$L = cVR\sqrt{2\rho T}$$

where the terms have the same meaning as in the text. Find the torque necessary to change the direction of the angular momentum if the axis of the flywheel is horizontal and the car is going around a curve of radius $R_r$ at a speed $v$. Evaluate this numerically for several different size flywheels and several sets of $R_r$ and $v$.

6. Suppose that a fiber flywheel has an energy storage of $1.2 \times 10^6$ ft-lb. Assume some reasonable geometry, and calculate the amount that the flywheel slows down for a single acceleration from 40 mph to 70 mph for a 100 slug car. Compute the initial and final overall gear ratios needed to couple the flywheel to the driving wheels.

7. How long would it take for a 10 hp electric motor to fully "charge up" a 500 lb fiber flywheel?

8. Suppose one is on a long trip. Clearly a flywheel will have to be "recharged" several times each day. Suppose that the maximum time a person is willing to spend on that process is 10 min. How many horsepower are needed to put maximum energy into a 500 lb fiber flywheel in 10 minutes?

# Unit 8
# Waste Disposal in Space

## INTRODUCTION

The average U.S. citizen produces about 5 pounds of solid waste a day. Landfill, incineration, and dumping at sea are common methods of disposal. Unfortunately there are problems with each of these. Suitable land for landfill operations is becoming scarce and remote in many locations. Unless incineration is controlled carefully, there may be intolerable air pollution. Dumping in the oceans pollutes the water and often results in trash being washed up on beaches.

The question of recycling is quite involved. The first problem is that many different materials are mixed together in trash and are difficult to separate. Even if people were to presort their trash it would not help much, since in many cases the products are already mixed, as in the case of insulation around copper wire and the many materials in an automobile. The second problem concerns the economics of recycling. One issue is the varying freight rates. For example it is cheaper to ship iron ore than scrap iron. Another main problem concerns the market for recycled material. A major effort to recover used paper, for example, would provide more reclaimed fibers than there is a demand for. Because fibers get shortened and degraded, some new pulp must always be used. In the face of the present low demand for recycled materials, it now costs more to sort and recycle trash than the value of the materials. The burning of trash for the energy content, or the conversion into gas or oil appears promising. See Reference 5.

Space is vast. Why not dump solid waste there? We examine some of the problems and energy required in this Unit. It may be that certain extremely toxic or hazardous wastes, such as some of the radioactive wastes from nuclear reactors, will be put into space.

## BASIC CONSIDERATIONS

Because of the force of gravity, it takes work to put an object into space. The force is conservative, so there is a potential energy function

$$E_p = -G\frac{m_1 m_2}{r}$$

where the symbols have their usual meanings. This expression is valid for point particles or outside of spherically symmetric objects, in which case $r$ is the distance between centers. Thus if we wanted an object 1000 km above the surface of the earth the correct value for $r$ would be

$$r = 1000 + r_e$$

$$= 1000 + 6378 = 7378 \text{ km}$$

where $r_e$ is the earth's radius.

The energy needed for an object to reach an altitude $h$ above the surface of the earth (ignoring air resistance and the rotation of the earth) is

$$\Delta E_p = E_p(h + r_e) - E_p(r_e)$$

$$= Gm_o m_e \left(\frac{1}{r_e} - \frac{1}{h+r_e}\right)$$

This energy will permit the object to just reach the altitude $h$ with zero kinetic energy, in which case it will fall back to earth.

To stay at an altitude $h$, the object must be moving in an orbit around the earth, which will be assumed circular. The force of gravity produces the required centripetal acceleration:

$$G \frac{m_e m_o}{r^2} = m_o \frac{v_o^2}{r}$$

This expression may be rearranged to obtain the kinetic energy needed for orbit:

$$E_k = \frac{1}{2} m_o v_o^2 = \frac{1}{2} G \frac{m_e m_o}{r}$$

$$= \frac{1}{2} G\, m_e m_o/(r_e + h)$$

Thus the energy needed to put an object into orbit about the earth at an altitude $h$ is:

$$E = E_k + \Delta E_p$$

$$= \frac{1}{2} G \frac{m_e m_o}{r_e + h} + G m_e m_o \left(\frac{1}{r_e} - \frac{1}{r_e + h}\right)$$

$$= \frac{G m_e m_o}{r_e} \left( 1 - \frac{1}{2(1+h/r_e)}\right) \qquad (1)$$

The actual energy will differ somewhat from this expression: air drag will increase the energy needed, and the effects of the rotation of the earth will either increase or decrease the answer depending on the direction of the orbit relative to the direction of the earth's rotation. There is also some energy needed for the essential mid-course trajectory change, without which the object would still return to earth. We will neglect these effects.

The energy per unit mass is obtained by dividing Eqn. 1 by $m_o$. We also substitute numerical values (mks units) at this time.

$$E/m_o = \frac{G m_e}{r_e} \left(1 - \frac{1}{2(1+h/r_e)}\right)$$

$$= 6.257 \times 10^7 \left(1 - \frac{1}{2+3.136 \times 10^{-7} h}\right) \qquad (2)$$

As a numerical example it requires

$$6.257 \times 10^7 \ (1 - 1/2.3136)$$

$$= 3.55 \times 10^7 \ \text{J}$$

to place 1 kg into orbit at an altitude of $h = 10^6$ m.

ROCKETRY

If the above energy is imparted by a rocket (which is the only practical way), the total energy expended is much larger than that given by Eqn. 2. This is because during the earlier stages of acceleration not only is the "payload" being accelerated but the rocket structure and the remainder of the fuel as well. It can be shown that the speed at shut down of a single stage rocket depends only on the exhaust speed of the fuel and the ratio of the initial mass to the final mass of the rocket:

$$v_{sd} = v_{ex} \ln(m_i/m_f) \tag{3}$$

where $v_{sd}$ is the speed at shut down and $v_{ex}$ is the exhaust speed. This result neglects air drag and pressure thrust, and assumes that the force of gravity may be neglected compared to the rocket's thrust, which is reasonably valid for smaller rockets. The above equation is derived in Reference 2, for example.

If the rocket has moved through a negligible fraction of the earth's radius during the burning stage, the required speed at shut down may be obtained from Eqn. 1:

$$\tfrac{1}{2}m_o v_{sd}^2 = E = \frac{Gm_e m_o}{r_e}\left(1 - \frac{1}{2(1+h/r_e)}\right)$$

or

$$v_{sd} = \left(\frac{2Gm_e}{r_e}\left(1 - \frac{1}{2(1+h/r_e)}\right)\right)^{1/2}$$

$$= 1.119\times10^4 \left(1 - \frac{1}{2(1+h/r_e)}\right)^{1/2} \tag{4}$$

As $h \to \infty$, this expression reduces to the standard "escape speed" result.

As an example, for an altitude of $10^6$ m, we obtain

$$v_{sd} = 8.432 \times 10^3 \text{ m/sec}$$

If the exhaust speed is 9600 ft/sec = 2926 m/sec, we have in this case

$$\ln (m_i/m_f) = v_{sd}/v_{ex}$$
$$= 2.882$$

or

$$m_i/m_f = e^{2.882} = 17.85$$

Thus about 94% of the original mass of the rocket must be fuel.

The energy expended is $H(m_i - m_f)$, where $H$ is the chemical energy released per kilogram. The payload is perhaps one half of the final mass, so the energy per unit mass of payload $(m_{pl})$ is

$$E/m_{pl} = 2\ E/m_f = 2H\left( \frac{m_i}{m_f} - 1 \right)$$

For an altitude of $10^6$ m, and a typical value of $H = 4.6 \times 10^7$ J/kg, we find

$$E/m_{pl} = 1.55 \times 10^9 \text{ J/kg}$$

Note that this number is almost 50 times larger than the energy calculated by Eqn. 2.

Multistaging of the rocket can reduce the energy requirements since in effect empty fuel containers (and hence unneeded mass) can be discarded on the way. In any event the energy required is much larger than chemical decomposition energies (see Prob. 1). On the other hand, radioactive isotopes are not affected by chemical processes. Only nuclear methods requiring about $10^6$ as much energy in general as chemical processes are effective. Thus orbiting radioactive wastes may be *energetically* attractive. There are some obvious hazards associated with this proposal.

The space around earth would rapidly become cluttered if this method of disposal were deemed worth the large energy expenditure required. Thus orbits about the sun come to mind. We can approximate the energy required for an independent orbit about the sun by setting $h = \infty$ in Eqn. 2. We obtain $6.257 \times 10^7$ J/kg. Again if rockets are used a much larger total energy must be expended. See Prob. 4. The minimum energy will be somewhat different from this since the gravitational force of the sun will aid somewhat. See Probs. 5 and 6. Some sort of a midcourse maneuver should be used to insure that the rocket will never intersect the earth's orbit again.

REFERENCES

1. S. Glasstone, *Sourcebook on the Space Sciences* (Van Nostrand, 1965)

2. F. W. Sears and M. W. Zemansky, *University Physics*, fourth edition (Addison-Wesley Publishing, Reading, Mass, 1970) p. 124

3. L. Hodges, *Environmental Pollution* (Holt, Rinehart, Winston, N.Y., 1973) Ch. 13

4. M. A. Benarde, *Our Precarious Habitat*, revised edition (W. W. Norton, N.Y., 1973) Ch. 9

5. T. H. Maugh II, Science 178, 599 (1972)

PROBLEMS

1. Compare the energy needed to chemically decompose a substance with the energy needed to put the substance into space. Assume the molecular weight to be 100 (needed for computing the number of molecules per kilogram) and that about 50 eV of energy is needed per molecule. ( 1 eV = $1.60 \times 10^{-19}$ J )

2. Why doesn't Eqn. 2 give zero energy for $h = 0$ ?

3. What is the minimum value for $h$ for a practical orbit? What factors must be considered?

4. Find the total energy needed for a rocket to place a payload of 1 kg into orbit for $h = 400$ miles and $h = \infty$.

5. Find the minimum energy needed for an object to escape from the earth and into an orbit around the sun. Ignore the orbital motion of the earth. Hint: find at what point the gravitational forces of the sun and earth are equal in magnitude and find the change in potential energy from the surface of the earth to that point, including the potential energy due to the sun.

6. Same as Prob. 5, except do not ignore orbital motions. Consider the question of angular momentum.

7. Show that a speed equal to the speed of the earth in

its orbit about the sun is sufficient for a rocket to reach the surface of the sun. (The sun's surface can be considered the ultimate garbage dump.)

8. What is the minimum energy needed for an object to reach the surface of the sun? The orbital motion of the earth and angular momentum must be considered.

9. Instead of sending wastes into the sun, it could be sent into outer space. How much energy is needed for this?

# Unit 9
# Tidal Power and Pumped Storage

## INTRODUCTION

In the face of summer brown-outs or even black-outs and winter fuel shortages, society needs to consider all possible energy sources. Air pollution considerations compound the energy problem. Our most abundant fossil fuel, coal, creates the worst air pollution, especially sulfur oxides and particulates. The lesser-polluting fossil fuels, oil and gas, have a quite limited supply in the U. S., perhaps only 30 or 40 years (see Appendix 3).

Nuclear energy is often held to be the solution. Yet there are considerable doubts about the disposal of the radioactive wastes from fission, and the controlled fusion process is not proven. See Units 26, 27 and 28. Hydroelectric power is about 25% developed now in the US. Further damming of rivers creates conflicts with esthetic values and water transportation. In addition, the reservoirs get filled with silt after about two centuries. Even if fully developed, hydropower would meet only about 10% of our present energy needs. The extent of the geothermal resources is not well known, but they are estimated to be limited. Geothermal power is also made less attractive by the environmental impact of the gas- and mineral-laden steam and hot water, and the large amount of waste heat due to the low temperatures (see Unit 14). Solar power apparently offers a reasonable solution. See Unit 22.

## TIDAL POWER

In this Unit we explore quantitatively the potential for tidal power. Clearly this is a long term resource. The energy is derived from the kinetic and potential energies of the earth-moon system, which has been around for several billion years. The tides are currently turning some of that energy into heat, as the waters crash against the shore and flow in and out of bays. The basic concept here is to turn some of the presently wasted tidal energy into electrical energy. The main question to answer is the size of this resource.

Most places in the world have two high tides and two low tides in a "tidal day" of roughly 25 hours ($8.92 \times 10^4$ sec more exactly). The idea is to catch the high tides by damming a natural or man-made bay. When the ocean is at low tide the water is released and turns a turbine, generating electricity. The gates are then closed. When the ocean is at high tide, water is admitted to the bay, again generating electricity, and so on. There are four times a tidal day when energy can be generated.

How much energy is available at each filling or emptying? We can calculate this by considering the gravitational potential energy of a filled bay relative to low tide in the ocean. We assume a basin with vertical sides for simplicity. See Prob. 3 for a more realistic shape. The surface area is represented by A, and the tidal range (the difference between high and low tide) by R. Gravitational potential energy for small changes in elevation is mgh. Clearly different parts of the stored water are at different heights. We thus consider a small layer at height y and of vertical thickness $\Delta y$ and mass $\Delta m$ (see the figure). The gravitational potential energy of that layer is

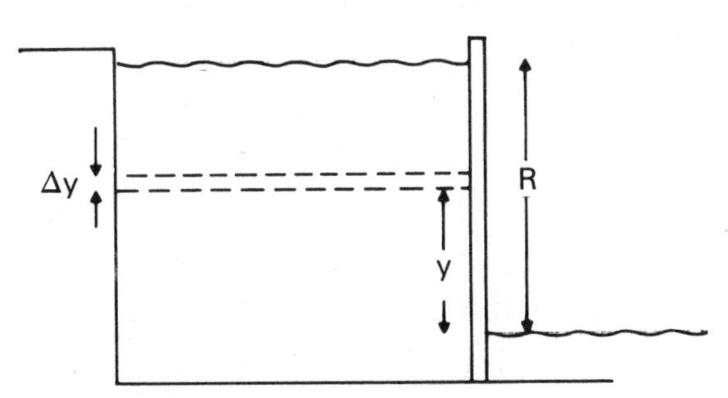


$$\Delta U = \Delta m y g$$

The mass of the layer is found from density (mass/unit volume), represented by $\rho$, and volume

$$\Delta m = \rho \Delta V$$
$$= \rho A \Delta y$$

Thus

$$\Delta U = \rho A g y \Delta y$$

The total potential energy is found by letting $\Delta y \rightarrow dy$ and integrating:

$$U = \int dU = \int_0^R \rho\ Agydy$$

Since we assumed vertical sides, $A$ is constant. Thus

$$U = \rho Ag \int_0^R y dy$$

or

$$U = \frac{1}{2} \rho AgR^2 \qquad (1)$$

The reader can show that the same energy is obtained on filling (see Prob. 2). In actuality the filling and emptying take a period of time so that the effective height difference is somewhat less than the maximum range.

As a numerical example let us assume that the bay is 3 km by 20 km, and that $R = 1.5$ m. The density of sea water is $1.025 \times 10^3$ kg/m$^3$. Thus we have

$$U = .5(1.025 \times 10^3)(3000)(20000)(9.8)(1.5)^2$$
$$= 6.78 \times 10^{11} \text{ J}$$

What average power results? Average power is energy/time. Since there are four cycles per tidal day of $8.92 \times 10^4$ sec, the average power is

$$\bar{P} = 4(6.78 \times 10^{11})/8.92 \times 10^4 = 3.04 \times 10^7 \text{ W}$$

The actual power output of a tidal electric plant is only 10 - 25% of the ideal given above. Hydroelectric power plants on the other hand are over 90% efficient. Problem 1 asks for various reasons why the efficiency is so low for tidal power. The electrical output in our example would therefore range between $3.0 \times 10^6$ W and $7.6 \times 10^6$W. The tidal electric plant at la Rance, France, has an average electrical power output of $6.2 \times 10^7$W. A modern fossil fuel plant has a power output of 100 to 1000 MW for comparison.

Because Eqn. 1 depends quadratically on R, places where there is a large range are especially favorable. For example the average range at la Rance is 8.4 m, and in places in the Bay of Fundy gets as high as 15 m. We will examine briefly some of the factors affecting the range. Out in mid-ocean the tidal range is on the order of 1 m. Along an open coast the water so to speak "piles up" at the shore and the range is more like 1.5 m. The tidal range varies with the phase of the moon. This is because while the moon is mainly responsible for the tides, the sun also produces a tide about 45% that of the moon's. See Prob. 6. When the sun and moon are in line, as at full moon or new moon, the effects add and the range is greatest. These are called spring tides. When the sun and moon form a right angle with the earth, as at first and third quarters, the effects oppose each other and smaller tides result, called neap tides.

The really tremendous ranges occur in bays, and are the result of resonance or standing wave behavior. The idea can be explained qualitatively rather easily. Consider a pan of water. If it is disturbed slightly, as by tilting it, the water will start sloshing back and forth. The time required for one cycle will depend on the size and shape. If one rocks the pan back and forth at the same rate as the sloshing rate, quite large amplitudes can be built up, just as a large amplitude on a swing can be built up from many small pushes all at the same rate as the swing's. A similar thing can happen in bays. Here the periodic driving force comes from the periodic rise and fall of the open ocean tides. Things are actually much more complicated, and depend upon whether the connection to the ocean is large or small, among many other things. See Reference 2.

The really important question to be answered before one can say whether tidal power could be an important resource, is how much power is available. There is a rather good estimate of the total rate of energy dissipation by the tidal flow, and it is about $3x10^{12}$W. The information comes from the slight lengthening of the day. Tidal friction causes the earth to slow down, and the power may be calculated from the rate of decrease in rotational kinetic energy of the earth. The rate of increase in the length of a day is about .00001 sec per year. This number has been established from relatively recent astronomical data and from ancient records of solar eclipses. It has been estimated that about one third of this, or $1x10^{12}$W, is in shallow seas and thus potentially available by building artificial basins. Since the conversion to electricity is about 20% efficient, only $2x10^{11}$W of electricity is available world wide from tidal power. This amount is about equal to one quarter of the present world electricity consumption. If one limits the capturing of tidal power to convenient natural bays, the resulting electric power is only about $1.5x10^{10}$ W. It is clear that tidal power is a minor source of energy.

There are some obvious drawbacks to tidal power besides the limited supply. The most obvious is that it is not continuously available. Some back up or energy storage must be provided. It can not be used for peak power demands, since the time of the tides shifts about an hour a day and does not necessarily coincide with the peak demand times. A two-basin approach can make power available as desired, but at a considerable loss in potential power. First of all, with two basins, each basin is only half the area that a single basin would be. Another reason is that the lower basin fills up as the higher one empties, so so that the effective tidal range is less. See Prob. 5. A large set of dams, especially if artificial basins were made, is quite unesthetic. Locks would have to be provided, and commercial and recreational boating would be impaired. The tidal ecosystem depends on the regular ebb and flow of the tides and would probably be upset.

## PUMPED STORAGE FACILITIES

A somewhat related concept is the use of two reservoirs at different heights for energy storage. The reason for considering energy storage is that now power plants must have sufficient generating capacity to meet the peak demand. Late at night the demand is much lower and much of the equipment is idle. Tying the power plants of the nation together helps some, because of the different time zones which shift the peak times slightly, but does not help the major day-night difference. The installed capacity is nearly twice the average power production. This means there is a considerably higher capital investment, and thus cost for electricity, than the average production requires. Nuclear power plants run much better at constant power than at varying power levels. With two reservoirs one can use the excess power capacity at night to pump water from the lower reservoir to the higher one. Then during high demand times the water can be used to run a hydroelectric generator. The losses are rather low. Solar power obviously also needs an energy storage method.

Generally the difference in elevation is sufficient that the changes in elevation as the upper basin empties can be neglected. The energy storage is thus just $mgh$ or $\rho Vgh$.

Unfortunately there are not many natural settings of existing lakes or even sites suitable for artificial reservoirs. Even when they do exist there is considerable opposition because of the esthetic damage. Thus pumped storage is not a major solution to the uneven demands for power.

REFERENCES

1. M. King Hubbert, "Energy Resources," in Resources and Man (W. H. Freeman, San Francisco, 1969)

2. E. P. Clancy, The Tides (Doubleday & Co, Garden City, N.Y., 1968) See especially Chs. 6 and 8.

PROBLEMS

1. Think of several factors which explain why the actual energy converted to electricity is only about 20% of the theoretical maximum.

2. Show that the energy available on filling a basin is the same as given by Eqn. 1.

3. Actual bays do not have vertical sides. A somewhat more realistic approach is to assume sloping sides as shown. Calculate the energy storage in this case.

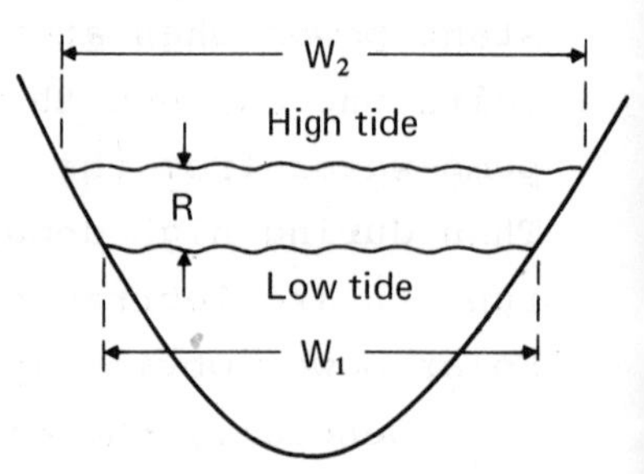


4. Assume that a power plant has a capacity of 500 MW, that the load during the night (8 hours) is only

250 MW, and that the average daytime power during non-peak hours is 500 MW. If there is an energy storage facility how much peak power can be provided if the peak period is 4 hours long? Calculate the size of a pumped storage facility if there is a 50 m vertical separation and a 5 m change in elevation of the reservoir is possible. Neglect losses.

5. How much does the stored energy decrease if double tidal basins are used for the purpose of shifting the time, instead of a single tidal basin dependent on the time of the tides?

6. The strength of the tidal force depends on the difference in the force of gravity between the side of the earth closest to the body in question and side of the earth away from the body. Show that although the sun is much more massive than the moon, its tidal force is only about half that of the moon.

7. Show that an increase in the length of the day by $10^{-5}$ sec/yr implies a loss in rotational kinetic energy of the earth at a rate of about $10^{12}$ W. Assume the earth is approximately a uniform sphere of density $5.5 \times 10^{3}$ kg/m$^3$. A more realistic calculation would be based on the actual variation of density with radius.

# Unit 10
# Noise Pollution

## INTRODUCTION

Defining "noise" presents some difficulties. Your high fidelity set may be enjoyment to you and annoyance to your neighbor. The sound of a fire engine is noise to the person awakened in the middle of the night, but an eagerly awaited signal to the family whose house is burning. Historically, many forms of music have been classified as noise at first by traditionalist critics. Perhaps the most satisfactory definition for our purposes is to say that noise is unwanted sound or excessive sound.

The most common aspect of noise is that it annoys. Annoyance is very difficult to quantify, since one can become conditioned to the source and "tune it out" to the extent that one no longer even notices it. There are psychological effects, such as the case where people were less bothered by airport noise after being told that an attempt to control it would be made. The level of annoyance of extraneous sound depends on the activity of the listener, such as trying to sleep, listening to quiet music, or having a loud party, and on the amount of masking by background noise.

The greatest harm from noise comes in the loss of hearing upon repeated exposure to loud noises. Certain industrial settings are very dangerous to hearing. It was common to call noise-induced hearing loss "boiler maker's ear". Deafness among the makers of copper pots was noted in Roman times!

Other important aspects of noise pollution include the interference to communication, primarily the spoken word, which is obviously important in offices and schools, and loss-of-privacy. The latter is actually a two-way noise problem, and is particularly acute in multifamily dwellings. Outside noise can intrude with obvious nuisance. Somewhat less often thought of, but equally important, is that if you can hear other poeple, they car hear you. People can not act naturally at home if they realize that their financial discussions, fights, and tender moments can be heard next door.

There is some suggestive data to the effect that excessive noise can affect health. Much of the evidence is based on laboratory animal experiments and circumstantial evidence with people. However, there is no general agreement on the existence or magnitude of the effect and whether continued exposure to the noise leads to satisfactory adaptation with no permanent results. There is a general stress-type reaction to short exposures to loud noises. For example heart rate increases, blood vessels constrict, and the digestion can be upset.

Still other effects include nervousness, irritability, hostile behavior, loss of concentration, and loss of sleep.

Since there is no simple relation between the noise and its effect in many of the above, we will consider in detail here only those for which there is a reasonably well established quantitative relationship. We must first define some of the units used to measure sound.

## QUANTIFYING SOUND

Frequency and duration are obvious characteristics of sound which need no further discussion here. The many units for measuring intensity must be understood. The fundamental physical units are intensity, which is sound power per unit area, and root-mean-square excess pressure. Because the ear is sensitive to a relative intensity range of about 1 to $10^{14}$, and because of the way the ear responds to small changes in intensity, a logarithmic scale has been found useful. Although the original unit was the Bel (named after Alexander Graham Bell), a unit one-tenth as large, the decibel (dB), is now in common use. Logarithms can be taken only of dimensionless quantities, so a ratio of intensity to some standard intensity, $I_o$, is used. Sound intensity level in decibels is given by

$$\text{dB} = 10 \log \frac{I}{I_o} \qquad (1)$$

The standard value of $I_o$ is $10^{-12}$ $\text{W/m}^2$. It is quite important to appreciate the logarithmic dependence. Going from 40 dB to 60 dB may not seem like much. Certainly going from 40 mph to 60 mph is not very significant. But going from 40 dB to 60 dB means a hundredfold increase in the intensity. It is also important to realize that 60 dB does not sound ten times louder to the ear than 50 dB. For sounds louder than about 40 dB the average judgment holds that a 10 dB increase in intensity sounds about twice as loud. See Problem 8.

Intensity is difficult to measure unless the sound is coming from only one direction, so the quantity usually used is based on pressure. Since for a single wave the intensity is proportional to the square of the rms excess pressure p, we define

$$dB_{SPL} = 20 \log \frac{p}{p_o} \tag{2}$$

The subscript SPL stands for "sound pressure level." Just as in sound intensity level (Eqn. 1), we make use of a ratio. Here $p_o = 2 \times 10^{-5}\ N/m^2$. For single waves the results of Eqns. 1 and 2 are virtually the same.

The reference levels in Eqns. 1 and 2 were chosen so that the minimum sound that the normal ear can hear in the region of 2000 Hz is 0 dB. Discomfort sets in at about 120 dB and real pain around 140 dB. The ear is most sensitive to sound in the 1000 to 5000 Hz region. At 50 Hz the minimum audible level is 40 $dB_{SPL}$ and at 10000 Hz, it is about 15 $dB_{SPL}$, for example.

Sound level meters usually have special networks built into them so that the meter's response is more nearly like that of the ear. The most commonly used weighting network is called "A", and matches the response of the ear at moderate levels rather well. Frequencies below 1000 Hz are de-emphasized considerably (at 50 Hz it amounts to 30 dB), and the frequencies about 5000 Hz are de-emphasized to small extent. The lesser-used "B" and "C" networks are more nearly uniform in response. When a weighting network is used, the sound level is reported as "70 dBA" for example. Some common sounds and the corresponding sound level are given in Table I, to establish a feel for the unit. One final refinement is that the meter can either have a "fast" response, or a "slow" response which averages over about a 1/2 sec period. Most often the unit used is "dBA-slow". For still other units used in noise measurement, see References 3, 4, and 5.

Table I. Some sound levels (A weighting)

| | |
|---|---|
| whisper | 30 dBA |
| quiet house | 40 - 45 dBA |
| average office | 50 - 60 dBA |
| dish washer, 3 feet | 65 dBA |
| vacuum cleaner | |
| operator's position | 80 dBA |
| 10 feet | 75 dBA |
| heavy street traffic | 80 - 90 dBA |
| electric lawn mower | |
| operator's position | 90 dBA |
| riveting machine | |
| at 35 feet | 100 dBA |
| jet airplane at | |
| takeoff, 200 feet | 120 -130 dBA |

HEARING LOSS

The U.S. Department of Health, Education and Welfare states: "the population at risk with regards to noise-induced hearing loss may be greater than any other hazard in the work environment." Only in the last quarter century have workers been able to collect damages for the gradual loss of hearing on noisy jobs. Legislation setting maximum permissible sound levels has come only in the last few years. The main law is the Occupational Safety and Health Act

(OSHA) of 1970. The current limit permitted depends on the exposure time, and is given in Table II. At no time can the noise level exceed

Table II. Present noise limits permitted under OSHA

| | | |
|---|---|---|
| 8 | hours/day | 90 dBA-slow |
| 4 | | 95 |
| 2 | | 100 |
| 1 | | 105 |
| 1/2 | | 110 |
| 1/4 | | 115 |

115 dBA. Since loud, very short duration noises do not show up on the slow scale, there is a separate limit of 140 dB peak for impact noises. The standards of Table II are proposed to be lowered by 5 dBA at some time in the future. It is important to note that present data show that the standards of Table II will still cause 20% to 30% more hearing impairment to workers after 20 years than for the general population. The standard would have to be dropped to 80 DBA for 8 hours to keep hearing loss minimal. This is not considered technologically feasible now.

If a worker is exposed to a variety of sound levels, the following formula must hold:

$$\sum_i \frac{ti}{T_i} \leq 1$$

where $t_i$ is the actual time spent at the $i^{th}$ level, and $T_i$ is the permitted time at that level. For example, a worker could be exposed to 90 dBA for 4 hours, and to 100 dBA for one hour under the law.

If the noise source is broadband (many frequencies

present), premanent hearing loss first appears at about 4000 Hz, gradually spreading to lower frequencies. If the noise source is narrow band, the loss of hearing for some reason first appears at about 1/2 octave higher. A hearing loss of 25 dB between 1000 and 3000 Hz is considered an impairment. These frequencies are important for speech. Some people are more susceptible to noise-induced hearing loss than others. Thus early testing is important to catch the hearing losses while they are still limited.

The above discussion concerns the permanent hearing loss caused by many years of exposure. There is also temporary hearing loss caused by short (a few hours) exposure. Anyone who as sat up front at a rock concert or drag race can testify to this point. Recovery may take around a day. The louder and longer the exposure, the longer the recovery. The recovery period is always much longer than the exposure period.

## SPEECH INTERFERENCE

The ability to communicate verbally is important for many jobs and social functions. It is common experience that noise makes speech hard to understand. The main frequencies of speech lie between 300 Hz and 3000 Hz. Every part of this frequency band is about equally important. Noise lying within this range is much more devastating to communication than noise lying outside this region. The concept of speech interference level (S.I.L.) is based on the foregoing fact. S.I.L. is defined as the arithmetic average of the sound pressure level in dB in each of the octave bands centered at 500, 1000, and 2000 Hz.

For normal conversation at a distance of 3 ft, the

S.I.L. must be less than 60 dB. If the S.I.L. is above 88 dB, one can not hear even a shout at one foot.

## NOISE CONTROL

One can divide control techniques by the location where the control is applied: at the source, along the transmission path, and at the hearer. The first is the most desirable, and the last is the least. We can only outline some of the methods briefly here. See References 3, 4, 6, and 7 for greater details.

In some cases it is not possible to control the noise at the source. For example the sound may be essential, such as radio, television, or band concerts. While the voluntary listeners may want to hear the sound, it could be noise to involuntary listeners. Another category includes the cases where technology has no answers at present, such as jet aircraft.

However, in many cases noise can be reduced considerably at the source. Machinery can be redesigned to reduce noise. For example large vibrating surfaces are excellent radiators of sound. The basic physical principle involved here is impedance matching. Vibrational motion of solids is not readily transferred to air because of the large difference in densities. A small surface, even with large amplitude vibrations, will not cause large acoustical waves in the air, because it interacts with only a very small mass of air. (One can not fan with a knitting needle.) A vibrating surface with a large area will set a larger mass of air in motion, resulting in a more intense sound wave. A simple example is a tuning fork. If it is held in the air it does not make much sound. If the butt of it is held against a table top, the intensity of the resulting sound wave is much greater. The bodies of violins and guitars and the sounding boards of pianos serve the same purpose. Thus often the noise radiated by a piece of machinery can be reduced by reducing the area of the vibrating surfaces, by making the surfaces stiffer to reduce amplitude, or by coating them with energy-absorbing material to dampen the vibration. Since large amplitudes result when the driving force is at the same frequency as a natural vibrational frequency, care must be exercised to avoid such resonances when regular periodic motion is present. It should be recalled that acoustical waves in air are longitudinal, so vibrational motion of a surface which is parallel to the surface will not cause appreciable sound.

Careful balancing of rotating parts will reduce vibration. Tire noises can be reduced by changing the tread design. If the tread design it too regular, a loud "singing" noise results. Trucks and construction machinery could have more effective mufflers without much loss of efficiency. Regular maintenance to keep things oiled, bolts tight, and worn parts replaced will reduce vibration and hence noise.

Frequently the best noise control at the source is to replace a noisy process with a quieter one. For example sometimes welding can replace riveting, cutters can replace jack-hammers, and pressing can replace pounding.

If the noise can not be stopped at the source, then attenuation along the path must be used. The first way to reduce the transmission of sound is to isolate the source if possible. This means not only surrounding it with a sound barrier, but also mounting it on isolation mounts. Otherwise the vibrations may be transmitted along the floor and by-pass the barrier. One common erroneous idea is that something like acoustical tile, which is effective for decreasing the reverberation of sound in a room, is a good sound barrier. This is not the case. The reason is that acoustical tile is not a sound absorber, but rather a poor reflector of sound. An open window is an example of something which reflects no sound and which transmits it all. In general the effectiveness of a barrier increases with the mass per unit area. Thus lead and concrete are more effective than lighter materials. As an approximate rule for single layer materials, every doubling of the mass per unit area results in an additional 5 dB decrease in the transmitted sound. See Problem 7. Some materials exhibit considerable deviation from this rule of thumb. For example, cinder blocks are rather poor attenuators. A 4 in. thick concrete slab will attenuate about 45 dB, while a 4 in. cinder block wall, which has about half the mass per unit area, attenuates only 25 dB.

Two thin layers separated by an air space are more effective than one single layer of the same total thickness. A 1/2 in. glass plate will attenuate about 30 dB, while two 1/4 in. glass plates will attenuate by 35 and 45 dB when separated by 1/2 in. and 6 in., respectively. In general any interruption of a transmission path will help considerably in reducing the transmitted sound, since the discontinuity causes a portion of the energy to be reflected. The general concept of impedance mismatch applies here. This is why a closed window reduces the intrusion of outside noise, even though glass is not a particularly good absorber.

When enclosing a noise source, one must be very careful to treat ventilating ducts and to seal the doors so as to prevent paths for sound to escape. Ventilating ducts are excellent carriers of sound. They act like wave guides, reflecting the sound from the walls repeatedly with little loss. Some sort of muffler must be provided in an air duct. Right angle bends help some.

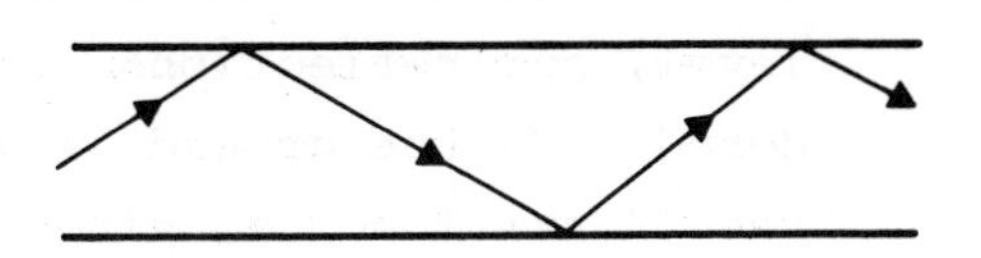

Outdoors increasing the distance between the source and the listener reduces the sound level. We can find the approximate rule for the decrease in sound level by applying conservation of energy. Consider first a point source in free space. If the absorption of sound by air is neglected, all of the power passing through a sphere of radius $r_1$ also passes through a sphere of radius $r_2$. Since intensity is power per unit area, we have

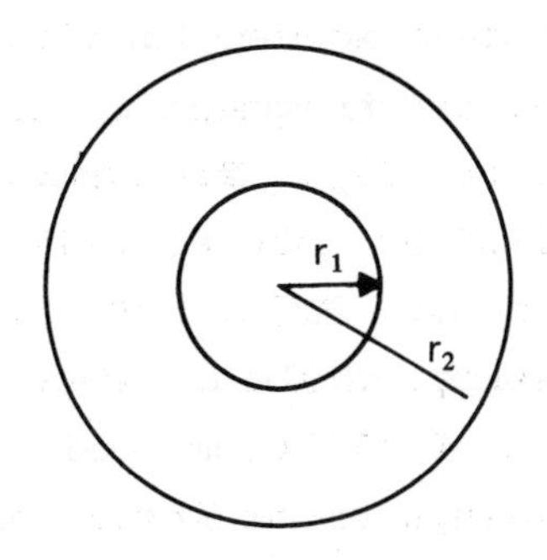


$$I_1\, 4\pi\, r_1^{\,2} = I_2\, 4\pi\, r_2^{\,2}$$

which leads to

$$\frac{I_1}{I_2} = \frac{r_2^{\,2}}{r_1^{\,2}}$$

Intensity thus varies inversely with the square of the distance. A doubling of distance reduces the intensity by a factor of four. Since the logarithm of 4 is .6021, the sound level in decibels is decreased by about 6 dB for each doubling of distance. The model of a point source in free space is almost never valid except for aircraft noise. Usually the source is near ground level, and reflections from the ground must be considered. If the ground is completely absorbing, the above result still holds, since the area of a hemisphere also varies as the square of the radius. If the ground reflects a significant portion of the sound, interference effects are present which are too varied to give a general result. The case of a line source is treated in Prob. 5. Air actually absorbs sound. The absorption is a complicated function of temperature, humidity, pressure, and frequency. High frequencies are absorbed more readily. As examples of the size of the effect, a 1000 Hz signal will be attenuated about 0.6 dB per 100 m at 20°C, 30% relative humidity, and sea level pressure, while a 4000 Hz signal will be attenuated about 4 dB/100 m under the same conditions. Inside buildings sound levels may sometimes be an increasing function of distance because of the presence of reflected waves.

If the source can not be surrounded with sound absorbing materials, one can try to prevent the sound from entering another room or building. It is obviously cheaper to treat one sound source if possible than to treat many rooms or buildings. The principles of preventing sound from entering a room are the same as above: build massive walls, floors, and ceilings, have double windows, and well-sealed doors. It is usually stated that 50 to 55 dB of isolation is desirable for privacy in a dwelling. There is little legislation on this point in the U.S., and many apartments do not meet this standard.

If nothing else can be done, the final line of defense against noise is ear plugs and acoustic ear mufflers. About 15 to 30 dB reduction is available by this method. Cotton stuffed in the ears makes only a modest difference.

REFERENCES

1. L. L. Beranek, "Noise" _Scientific American_, Dec. 1966.

2. T. Berland, _The Fight For Quiet_ (Prentice-Hall, 1970). The first chapter contains several errors and is best ignored.

3. A. P. G. Peterson and E. E. Gross, Jr., _Handbook of Noise Measurement_ (General Radio Co., Concord, Mass).

4. L. L. Beranek (editor), _Noise and Vibration Control_ (McGraw-Hill, 1971).

5. K. D. Kryter, _The Effects of Noise on Man_ (Academic Press, 1970) Especially Chs. 2 and 9.

6. T. P. Yin, "The Control of Vibration and Noise" _Scientific American_, Jan. 1969.

7. R. Taylor, _Noise_ (Penguin Books, 1970).

PROBLEMS

1. How much more intense is 90 dB than a) 70 dB, b) 65 dB, c) 47 dB?

2. What is the increase in decibel level when the intensity of a source a) doubles, b) triples?

3. If two independent sources each separately have a noise level of 70 dB, they will have a noise level of 73 dB when together. Explain.

4. A very loud high fidelity set may have a sound level of 95 dB. How many watts per square meter is this?

5. How will sound level (decibels) vary with distance from a line source, such as a noisy road? Hint: sound radiates in a cylindrical pattern from a line source.

6. Find out whether the following noise combinations exceed the legal limit:
   a) 2 hours at 90 dBA, and 2 hours at 95 dBA, and 1 hour at 100 dBA.
   b) 4 hours at 90 dBA, and 1/2 hours at 105 dBA.

7. Show that doubling the mass per unit area leads to an approximate decrease in the transmitted sound by 6 dB. Hints: sound pressure (force/unit area) provides the force that moves the wall. What happens to the amplitude of the acceleration when the mass per unit area doubles? How is the resulting displacement altered? How does intensity of sound on the other side relate to the displacement? How is intensity related to dB? The value of 5 dB for each doubling given in the text more nearly represents the actual case.

8. Several physics texts erroneously claim that the decibel is used because the ear hears in a logarithmic fashion. Perceived loudness actually is approximately proportional to the rms excess pressure raised to the 0.6 power. Find the change in loudness as one goes from 40 dB to a) 50 dB, b) 60 dB, and c) 80 dB. If we heard logarithmically, the last increase would be only a doubling in loudness.

# Unit 11
# Shock Waves and the SST

## INTRODUCTION

While any object travelling through a medium faster than the appropriate wave speed will create a shock wave, the main environmental concern is with acoustical shock waves or "sonic booms" created by aircraft travelling faster than sound. At the present, the US effort to build a commercial supersonic transport, or SST, has been halted. However, it could be revived, and, in any event, other countries are proceeding with development. Not only is there the sonic boom problem with flights over land, but the SST also raises an energy question and might produce a possible change in climate.

It is estimated that the SST will consume about 12 gallons of fuel per mile, thus getting only .08 miles per gallon. Further, it would carry only relatively few passengers. If there are 200 passengers, the passenger-mile/gallon figure is only 16. So it uses more fuel per passenger than a 747, which obtains around 25 passenger-miles per gallon, and is about as bad as a large automobile carrying two people, or a smaller one with one person. Trains, for comparison, can achieve around 100 passenger-miles per gallon. So in a country becoming more concerned about energy conservation, the SST does not make sense.

It is generally conceded that a few SST's flying in the stratosphere will not create a noticeable change in the climate. It is unclear what many would do. Some experts feel that the vapor trails and particulate matter injected into the stratosphere might trigger a change in climate, since the stratosphere is vertically stable, and matter remains there a long time, thus allowing a considerable build-up. It is generally held that it would take far more SST's than are likely to fly to make a significant effect, however.

In addition to the sonic boom problem, there is enhanced subsonic noise pollution at and near the airport. Take-off and landing speeds will be higher, since the proper wing shape for supersonic flight gives relatively little low speed lift. Thus, engines with higher thrust are needed (see Prob. 5), which means greater noise. Some estimates indicate perhaps 10 to 20 db more noise at take-off. Other problems include additional radiation exposure from cosmic rays, because the SST flies higher, more danger from a sudden de-pressurization of the cabin, and greater difficulties in controlling the aircraft, due to the long length, high speeds, and small control surfaces. See Reference 1 for more details on the above.

## SHOCK WAVES

The major problem is the shock wave which is created continuously whenever the flight speed is greater than the speed of sound. Note carefully that a shock wave is created all the time, and not just when the speed of sound is first attained, as is sometimes erroneously thought. It is convenient to express speeds as a ratio of the speed, $v$, to the speed of sound, $c$. This ratio is called the Mach number:

$$M = v/c$$

$M$ is less than one for subsonic flight and greater than one for supersonic speeds.

A very pictorial way to think of the formation of shock waves is the following. Normally, with $M < 1$, the coming presence of an object is signalled ahead by a compression wave travelling at the speed of sound. When $M > 1$, the object is there "before the air knows it is coming." A much more satisfactory way is to consider it a case of constructive interference. When an object is travelling less than the speed of sound, the wave fronts emitted at various times remain distinct. See Fig. 1. However, when $M$ is greater than one, the successive wave fronts all have a common envelope, where constructive interference occurs. See Fig. 2. This envelope is the shock wave. We can find the angle of the shock wave cone as follows. Consider a wave front emitted when the object was at 0 in Fig. 3. During a time $\Delta t$, the front expands as shown, with a radius $\overline{OB} = c\,\Delta t$, where $c$ is the speed of sound.

During the same time $\Delta t$, the object will have moved to A. Thus, $\overline{OA} = v\ \Delta t$. The angle $\theta$ is found from

$$\sin\theta = \overline{OB}/\overline{OA} = c\ \Delta t/v\Delta t = c/v$$

$$= 1/M \tag{1}$$

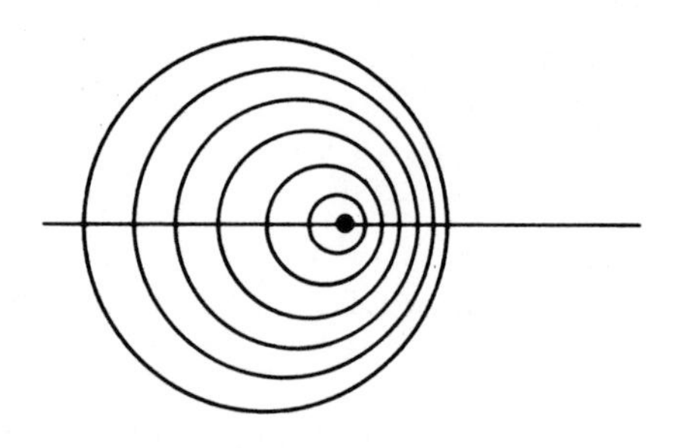

Fig. 1 An object moving less than the speed of sound, showing expanding wave fronts centered on successive positions.

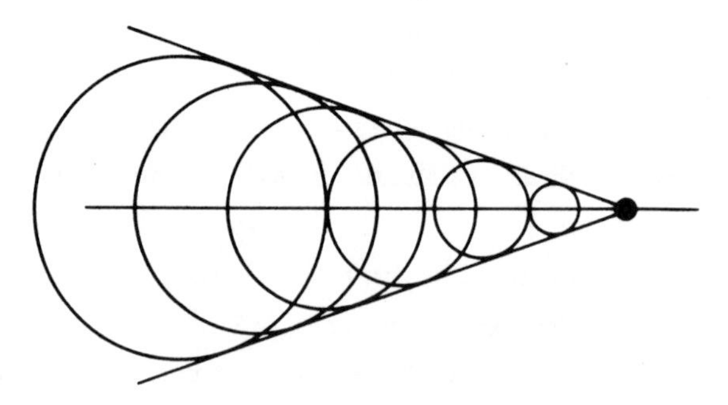

Fig. 2 An object moving faster than the speed of sound creates a shock wave. Since there is rotational symmetry about the direction of travel for a point source, the shock wave is a cone.

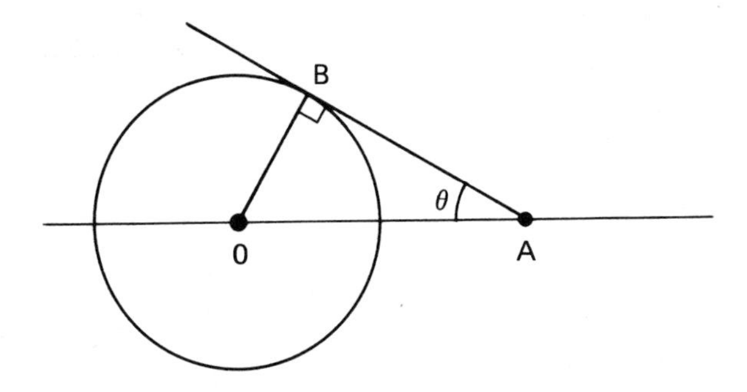


Fig. 3 Diagram for finding the angle of the cone.

We see that the higher the Mach number, the narrower the cone.

A shock wave is inescapably formed whenever $M > 1$ and is felt when it reaches the ground. The intersection of a cone and a plane parallel to its axis is a hyperbola. Thus, at any given time, people along a hyperbolic path are experiencing a "sonic boom." The path tends to be wider for higher speeds and high altitudes, and narrower for smaller $M$ and lower altitudes. Typical widths are 30 miles for an altitude of 3.5 miles and 50 miles for an altitude of 9 miles. Because an airplane is not a symmetric object, the intensity in different directions about the line of travel is not uniform.

In most cases of aircraft, the sonic boom is actually perceived to be double. This is because an overpressure wave is formed at the nose, and an underpressure wave at the tail. The typical change in pressure as a function of time is shown in Fig. 4. Each sudden rise is perceived as a boom. The change in pressure ranges from less than $0.3\ \mathrm{lb/ft^2}$, which is not noticeable, to over $5\ \mathrm{lb/ft^2}$ which causes structural damage. We can estimate the time between the two sudden rises by knowing that the speed of sound is about 1000 ft/sec, and the length of an airplane is around 100 ft to 300 ft. We obtain thus about 0.1 - 0.3 sec for the time between the two booms. In some cases there is deviation from the "N" shape shown in Fig. 4.

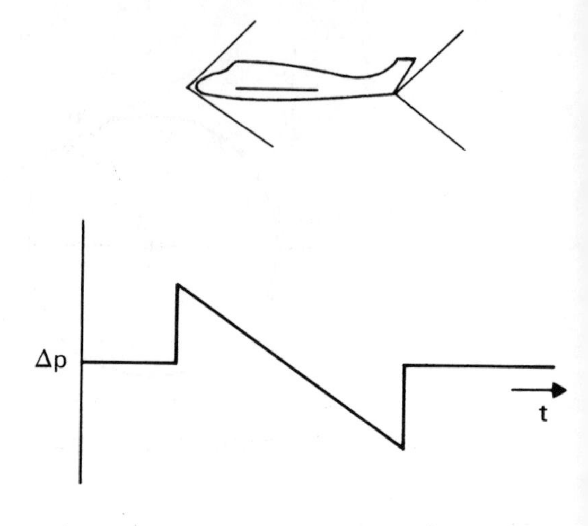


Fig. 4 The time variation in pressure in a sonic boom.

The change in pressure decreases only gradually as the altitude of the plane increases, going approximately as the inverse three-fourths power:

$$\Delta p \propto d^{-3/4} \tag{2}$$

Thus, every doubling of altitude reduces the severity of the pressure increase by about 0.6. This is a far cry from an inverse square law! Thus, little is to be gained by flying higher. While the intensity decreases some, the width increases at the same time. So a higher flight means more people receive a slightly less intense shock.

The variation of $\Delta p$ with speed is

$$\Delta p \propto (M^2 - 1)^{1/8} \tag{3}$$

where $M$ is the Mach number. Thus, for $M$ between 1 and about 1.4, the pressure change increases rapidly with speed. Above about $M = 1.4$ the pressure change increases only very slowly. See Prob. 4. One therefore need not worry about the sonic boom caused by barely supersonic flight. On the other hand, there is no advantage to $M$ slightly larger than 1 as compared to $M$ somewhat smaller than one. In general, the bigger the airplane, the bigger the excess pressure. The proposed US SST would have produced excess pressures about double those created by military fighter planes.

## SUPERBOOMS

Occasionally, a boom may have 2 or 3 times the usual excess pressure. The main causes are focusing by the atmosphere, superposition of a direct and a reflected wave, or the result of turns or accelerating. The air is not a uniform medium. Not only are there variations in pressure and temperature with altitude, but also horizontal variations. The speed of sound does not depend significantly on pressure, but is quite temperature dependent, increasing as the square root of the absolute temperature. An increase in temperature increases the wave speed, causing a refraction or bending of the wave front. This can focus some of the shock wave so that one point may have the shock wave approaching from several directions. See Fig. 5. Turning can also produce a similar effect. See Fig. 6. When an aircraft just achieves supersonic flight, the change in pressure is momentarily extra large.

## EFFECTS

People are definitely startled and annoyed when a sonic boom hits them. The reaction may be a mild disturbance, or a rather severe reaction. Sick people, especially those with heart problems, and nervous people are particularly affected. The sleep of millions could be

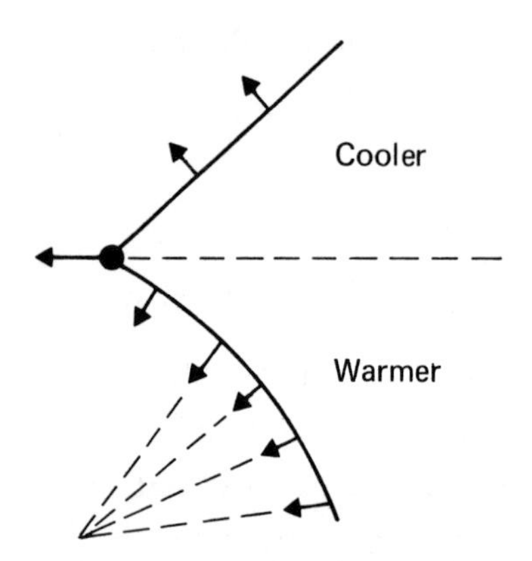


Fig. 5 Focusing of a shock wave due to an increase in temperature.

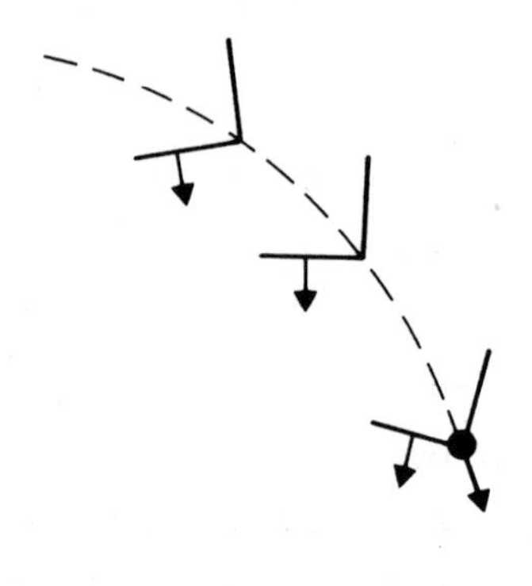

Fig. 6 Focusing due to turning.

affected many times a night if supersonic travel becomes commonplace. There is conflicting evidence about whether people can really adjust to sonic booms. If they are mild enough and come during the day on a predictable schedule, then the annoyance diminishes with continued exposure. On the other hand, it seems that unexpected sonic booms, and especially superbooms, will always produce an unfavorable reaction. The startling effect is often worse inside a building than outdoors, due to repeated reflections from the walls.

There are many cases of well documented structural damage due to sonic booms. The onset of damage occurs at an overpressure of around 5 $lb/ft^2$. Typical damage is minor: broken windows, cracked plaster and masonry, objects rattled off shelves, and the triggering of avalanches and landslides. In most cases, it can be argued that poor design was involved or that stresses had gradually built up over a period of time, and that the shock wave triggered the stress relief. Frequently, a house will suffer no damage from many sonic booms, and then for no apparent reason may be damaged by the next one. The fact that buildings *could* be built so as to not be damaged is of little comfort to the millions of owners of existing buildings. An overpressure of at least 10 $lb/ft^2$ is needed to cause damage in a well-built structure. It is definitely the case that many far-fetched and even fraudulent claims have been filed for alleged damage caused by sonic booms. A reasonably unbiased estimate places the damage at about $1000 to $5000 per flight over the US as an average.

## REFERENCES

1. W. A. Shurcliff, *SST and Sonic Boom Handbook*, (Ballantine, 1970).

2. H. H. Hubbard, "Sonic Booms," Physics Today 21 #2, 31 (Feb., 1968).

3. F. G. Finger and R. M. McInturff, "Meteorology and the Supersonic Transport," Science 167, 16 (Jan. 2, 1970).

4. K. D. Kryter, "Sonic Booms from Supersonic Transport," Science 163, 359 (Jan. 24, 1969).

PROBLEMS

1. Express $2\ \mathrm{lb/ft^2}$ in $\mathrm{N/m^2}$ and as a sound pressure level ($\mathrm{dB_{SPL}}$ with $p_o = 2\times10^{-5}\ \mathrm{N/m^2}$; see Unit 10).

2. It is well known that air pressure decreases as altitude increases. Find the altitude change which is equivalent to a pressure change of $2\ \mathrm{lb/ft^2}$. Hint: recall that the pressure at any level must be sufficient to support the weight of all the air above. The answer to Prob. 1 makes an excess pressure of $2\ \mathrm{lb/ft^2}$ appear large. The answer to the present problem makes it seem minor. The reason for the apparent discrepancy is that in a sonic boom the excess pressure is applied very quickly.

3. How much less excess pressure is there for an SST flying at a) 30 000 ft and b) 60 000 ft as compared to one flying at 20 000 ft?

4. Make a graph of the relative excess pressure as a function of Mach number.

5. How much thrust must each of the four engines have if a 750 000 lb plane is to reach 220 mph in 10 000 ft? Assume uniformly accelerated motion.

# Unit 12
# Air Pollution II: Thermal-gradient Particle Precipitators

The adverse effects of particulate matter in the air and some control methods were covered in Unit 4. In this Unit we consider another method whereby particles can be removed from the air. If a hot surface and a cold surface are close to each other, dirt will be deposited on the cold surface. An example which may be familiar to many is the case of the old-fashioned free-standing radiator located by an outside wall. Dust accumulates on the wall behind the radiator more rapidly than other places. The effect is more noticeable if it is a steam radiator, but still present with a hot water one.

The principle behind the thermal gradient precipitation is quite simple. According to the kinetic theory of gases, air is composed of molecules in rapid random motion. The average speed of the molecules is related to the temperature by

$$\bar{v} = \sqrt{\frac{8kT}{\pi m}} \tag{1}$$

where $k$ is Boltzmann's constant, $T$ the temperature on the Kelvin scale, and $m$ the mass of each molecule. We suppose that there is a uniform temperature gradient between the surfaces. Particles of concern are much larger than the size of molecules. Thus, they will be moving much more slowly than the molecules, and will have frequent collisions with the molecules. If the colder surface is on the left as shown in the Figure, the molecules coming from the left and hitting the particle will on the average be moving slower than those hitting it on the right. So more momentum directed to the left than to the right is transferred to the particle, and it moves to the left.

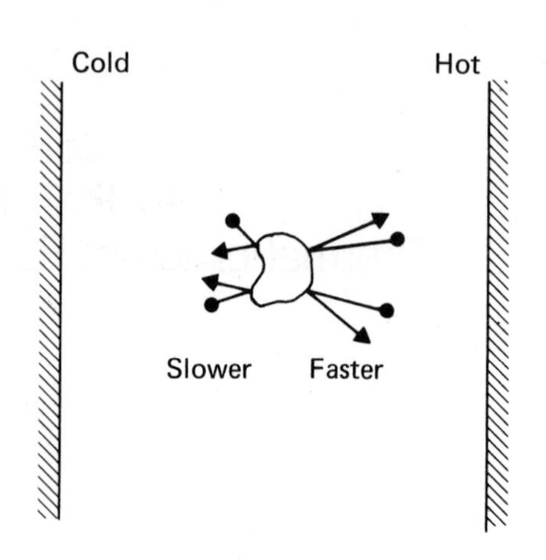


We can obtain an approximate expression for the thermal force on the particle from dimensional analysis. Intuitively, we would expect the force to be proportional to how rapidly temperature changes with position. This is called the temperature gradient, and, in the case where temperature is a function of one variable only, is just the derivative, $\frac{dT}{dx}$. We know that multiplying temperature $(T)$ by Boltzmann's constant $(k)$ gives a term with the dimensions of energy $(kT)$, and that force times distance also is dimensionally energy. Thus, we get the dimensions of force from Boltzmann's constant times temperature divided by length. Since a thermal gradient is dimensionally temperature divided by length, we obtain

$$F_{thermal} \propto k \frac{dT}{dx} \tag{2}$$

Dimensional analysis of course does not "prove" an answer, since there are many other combinations involving $\frac{dT}{dx}$ which have the dimensions of force. More complicated (and better) expressions are given in Ref. 1, Eqns. 1, 5, and 6. However, under certain simplifying assumptions these reduce to some dimensionless ratios times $k \frac{dT}{dx}$. The constant of proportionality depends on the size of the particle, and may range from perhaps $10^2$ to $10^8$ for particle sizes of interest.

The principal application of thermal gradient precipitators at present is for the collection of particles for air quality monitoring. Usually there is a microscope slide in contact with the colder surface, and either a hot wire or hot surface is used. The temperature gradient should be at least 750 °C/cm. The rate of gas flow must be very low, on the order of 10 cm$^3$/min, for good collection efficiencies. Efficiencies approach 100% for particles smaller than about 10 microns on down to molecular sizes under these conditions. Samples suitable for examination by an electron microscope are readily obtained. One major advantage of the thermal precipitator over filters and inertial (impaction) collectors is that the particle size is not affected by the collection. In the case of filters, particles tend to stick together, and in the case of inertial collectors, the particles tend to be broken into smaller sizes.

The principle of thermal deposition has not been applied to large scale pollution control. The energy needed to maintain the thermal gradient is about an order of magnitude larger than for other forms of air purification.

## REFERENCES

1. J. A. Gieseke, "Thermal Deposition of Aerosols," in _Air Pollution Control, Part II_, W. Strauss, editor. (Wiley-Interscience, 1972).

2. R. D. Cadle, _et. al._, "Sampling Procedures," in _Air Pollution Handbook_, P. L. Magill, _et. al._, editors. (McGraw-Hill, 1956) pp. 10 - 30.

## PROBLEMS

1. Suppose the hot and cold surfaces are separated by .5 cm, have an area of 1 $m^2$ each, and are maintained at a temperature difference of 1000 °C. Look up the thermal conductivity of air, and calculate the power needed to maintain the temperature difference. Does your answer change if the spacing is altered while maintaining the same temperature gradient of 2000 °C/cm?

2. Compare the thermal force (Eqn. 2, with the constant of proportionality taken as $10^7$) with the force of gravity for a particle 1 micron in diameter.

Assume the density of the particle is 5 $gm/cm^3$ and the thermal gradient is 2000 °C/cm.

# Unit 13
# Air Pollution III:
# Temperature Inversions

## INTRODUCTION

The level of pollutants in the air depends on both the rate of emission *and* the rate of removal or dilution. We discuss the latter here. Many factors affect the removal rate of pollutants: the rate of falling in the case of particles (see Unit 4), washout by rain, the rate of consumption in chemical reactions, dispersal and dilution by wind, conversion by microorganisms, absorption by plants, the ground and water, and dilution by vertical mixing. We consider the last factor in detail in this Unit.

Most air pollutants are emitted close to the ground. Clearly, if they stay there, the concentration of pollution will be higher than if they are mixed upwards into clean air. Wind which is deflected vertically by hills or structures or which is flowing turbulently will produce vertical mixing. Another important mechanism for vertical mixing depends on the buoyancy of air which is less dense than its surroundings. Since all air at a given altitude is at the same pressure (on a local scale), the density depends on the temperature: the hotter the air, the less dense it is. A given portion of air may be hotter than the surrounding air by having been emitted at an elevated temperature. However, a more significant case of vertical mixing occurs when the vertical temperature profile leads to general atmospheric instability.

## STABILITY AND INSTABILITY

The basic test for stability of any system involves the behavior of the system following a small displacement, real or imagined. If forces arise which act in a direction to return the system to its original position, the system is stable. An example is a ball at the bottom of a bowl. If the forces which arise as a result of the displacement act to increase the displacement, the system is unstable.

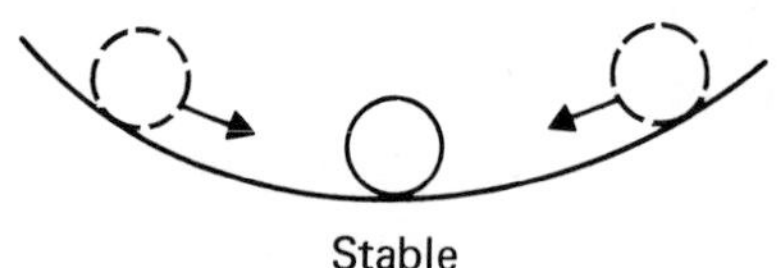

Stable

We can apply this test to check the vertical stability of air. We first note that pressure decreases as height increases. This is because each "layer" of air must support the weight of only the air above it. We now imagine moving a test volume of air upwards a small amount. Since the pressure

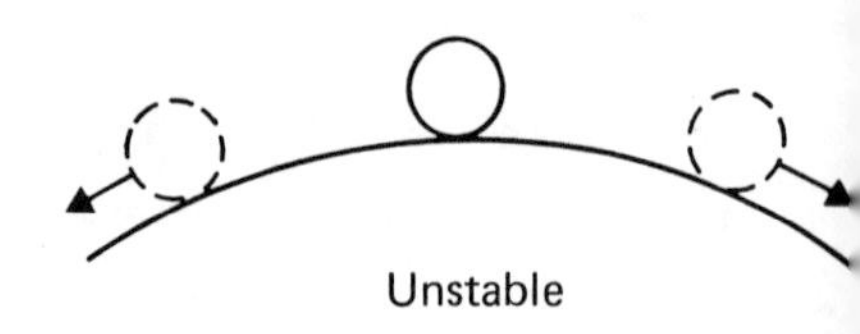

Unstable

decreases, the volume of the test sample increases. As the sample expands, its temperature falls. We imagine the test displacement to occur quickly enough that there is no exchange of heat. Thus the expansion is adiabatic. If our expanded-and-cooled sample is at a lower temperature than the surroundings, it will be denser and will sink back down. The air, in this case, is vertically stable. If, on the other hand, it is warmer than the surrounding air after being displaced, it will continue to rise, and the air is unstable. The reader should apply the above ideas to the case of a test displacement initially downwards.

## DRY ADIABATIC LAPSE RATE

We now seek to make the foregoing ideas quantitative. The term "lapse rate" means the rate of <u>decrease</u> of temperature with altitude $h$:

$$\text{Lapse rate} = -\frac{dT}{dh} \tag{1}$$

The minus sign is introduced since normally temperature decreases with altitude. Under such circumstances, the lapse rate as defined is positive. Since the temperature change with altitude is rather small, the lapse rate is often quoted in terms of a larger-than-unit distance, with temperature change per 100 m or per 1000 ft common. Dry adiabatic lapse rate refers to the rate of change of temperature with height for dry air which has been displaced adiabatically.

We start from the standard result for the change in pressure for a fluid in hydrostatic equilibrium:

$$dp = -\rho g \, dh \tag{2}$$

where $\rho$ is the density, $g$ the acceleration due to gravity, and $dh$ is the increment in height. We can obtain an expression for $\rho$ from the ideal gas law and the definition of density:

$$pV = \mu RT$$

$$\rho = m/V$$

where $\mu$ is the number of moles. The mass in a given volume is equal to the number of moles in the volume times the mass per mole, which is the same thing as the molecular "weight," $M$:

$$m = \mu M$$

Putting the last three equations together we obtain:

$$\rho = Mp/RT$$

Thus, Eqn. 2 becomes

$$dp = -(gMp/RT) \, dh$$

or

$$dp/p = -(gM/RT) \, dh \tag{3}$$

We now need to use the fact that we are displacing the test volume adiabatically. The usual result for the adiabatic expansion of gases is:

$$pV^{\gamma} = \text{constant}$$

where $\gamma$ is the ratio of the specific heats at contant pressure and constant volume. We need a relation between pressure and temperature, which can be obtained from the above equation by substituting for $V$ from the ideal gas law. The result is:

$$p^{\gamma-1} = k\, T^{\gamma} \qquad (4)$$

where $k$ is a constant. Since Eqn. 3 involves differentials, we take the differential of Eqn. 4 to obtain

$$(\gamma-1)p^{\gamma-2}\, dp = k\gamma T^{\gamma-1}\, dT$$

If we divide this equation by Eqn. 4 we obtain:

$$(\gamma-1)\ dp/p = \gamma\ dT/T$$

or

$$\frac{dp}{p} = \frac{\gamma}{\gamma-1}\ \frac{dT}{T} \qquad (5)$$

We can now combine Eqns. 3 and 5 to arrive at the desired result:

$$dT = -\frac{\gamma-1}{\gamma}\,\frac{gM}{R}\, dh \qquad (6)$$

The dry adiabatic lapse rate is thus:

$$-\frac{dT}{dh} = \frac{\gamma-1}{\gamma}\,\frac{gM}{R} \qquad (7)$$

which depends only on the kind of gas and $g$. Since the height of the atmospheric regions of interest in air pollution is small compared to the radius of the earth,

$g$ may be considered constant. An alternative derivation of Eqn. 7 is given in Ref. 3.

For dry air, $\gamma = 1.41$ and $M = 28.96$. For a kilogram mole, $R = 8.31 \times 10^3$. The numerical value of the dry adiabatic lapse rate is thus:

$$-\frac{dT}{dh} = \frac{(.41)(9.8)(28.96)}{(1.41)(8.31 \times 10^3)} = 9.9 \times 10^{-3} \text{ °C/m}$$

This value is very closely 1 °C for every 100 m or 5.5 °F per 1000 ft.

So far we have restricted ourselves to dry air. If the air is moist (which usually is the case, although it can be neglected in some arid conditions) we have to consider the relative humidity and ultimate condensation of the moisture into drops. Absolute humidity refers to the amount of water vapor per unit of air. Relative humidity compares the amount of water vapor present to the maximum amount or saturation level of water vapor at that temperature. Air can hold less water vapor at low temperatures than at higher temperatures. Thus as a given amount of air cools down the absolute humidity remains constant while the relative humidity increases until 100% relative humidity is reached. At that point if there are condensation nuclei present, water drops will form and grow. When water condenses from gas to liquid, the latent heat of vaporization is given off, which warms the surroundings. Thus a qualitative picture of the adiabatic wet lapse rate is as follows. As a given parcel of wet air goes up, its temperature falls approximately the same as for dry air (the effective value of $\gamma$ and $M$ change some with humidity) until 100% relative humidity is reached. Above this altitude, sufficient water vapor condenses to maintain 100% relative humidity as the temperature drops further. However, because of the latent heat released, the lapse rate is now less than for day air. No single value can be given for the wet lapse rate, since the saturation level depends on temperature. The wet lapse rate may be as low as .0035 ° C/m. A global average lapse rate is about .0065 ° C/m.

## STABILITY AND LAPSE RATES

We now put the two previous sections together. The controlling factor for the vertical stability of air is how the actual prevailing lapse rate compares to the abiabatic lapse rate. We consider here only dry air for simplicity. Figure 1(a) shows a case where the actual lapse rate is larger than the adiabatic case. This means that the temperature falls more rapidly than 1°C for every 100m. Consider a parcel of air at height $h_o$ and draw the dry adiabatic lapse curve through that point as shown by the dashed line in Fig. 1(a). If this test parcel of air is displaced vertically to height $h_1$ its temperature will follow the adiabatic relation. Thus it will be warmer than its surroundings at $h_1$. As we saw earlier, a parcel of air that is warmer than its surroundings will be less dense and will continue to rise. We therefore conclude that lapse rates larger than adiabatic result in air that is vertically unstable. This is desirable from an air quality viewpoint, since there will be good vertical mixing and dilution of the pollutants.

If the actual lapse rate is less than adiabatic, stability results. Refer to Fig. 1(b) where again the adiabatic behavior is shown as a dashed line. In this case when the test quantity of air is displaced

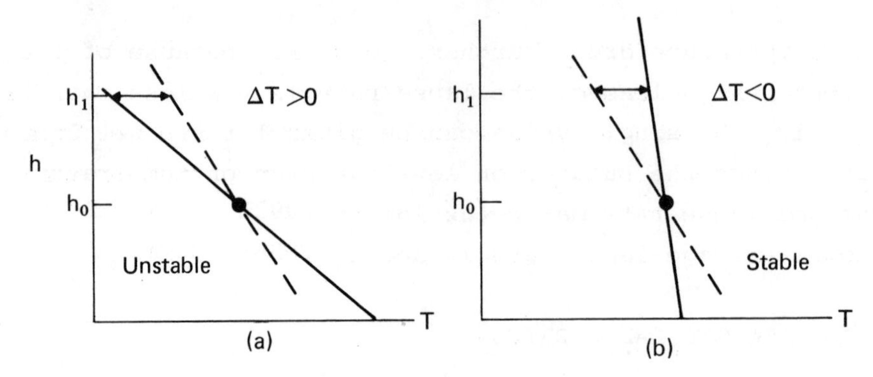


Fig. 1. The actual prevailing atmospheric lapse rates are shown as solid lines and the adiabatic lapse rate as dashed.

from $h_o$ to $h_1$ it will be cooler than its surroundings at $h_1$. Thus it is more dense and will sink back down, and we have a stable situation. We summarize these results graphically in Fig. 2. Wet air behaves qualitatively the same; the only difference is that the moist air lapse rate must be used.

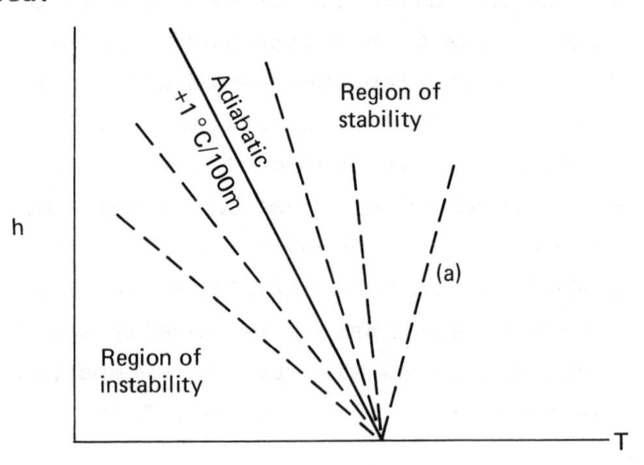


Fig. 2. Stability and lapse rates for dry air. The dashed lines are possible atmospheric lapse rates. Line (a) represents an inversion.

## INVERSIONS

Usually the temperature in the lower parts of the atmosphere decreases with altitude. Because of the way lapse rate is defined, this situation corresponds to a positive lapse rate. Sometimes the temperature increases with altitude, such as shown by line (a) in Fig. 2, leading to a negative lapse rate. Because this behavior is contrary to the usual one, it is called a temperature inversion. The concepts of "stable air" and "temperature inversion" are often used as if they were identical. While it is true that a temperature inversion means the air is stable, air can be stable for a positive lapse rate as well. The severest levels of air pollution are in the cases of inversions, however, since the stability is greater and the air is less likely to be disturbed by wind. Furthermore much of the polluted air is emitted at an elevated temperature and thus will rise at least for a while. The height to which it can rise is less when there is an inversion. See Problem 2.

Inversions can be of serveral forms. Mostly we need to consider only whether an inversion reaches the ground as in line (a) of Fig. 3, or whether it occurs at a higher altitude as in (b). Usually above some altitude the lapse rate becomes positive again as shown by the dashed lines in Fig. 3. Thus there is only a layer of air which has a temperature inversion; hence the term inversion

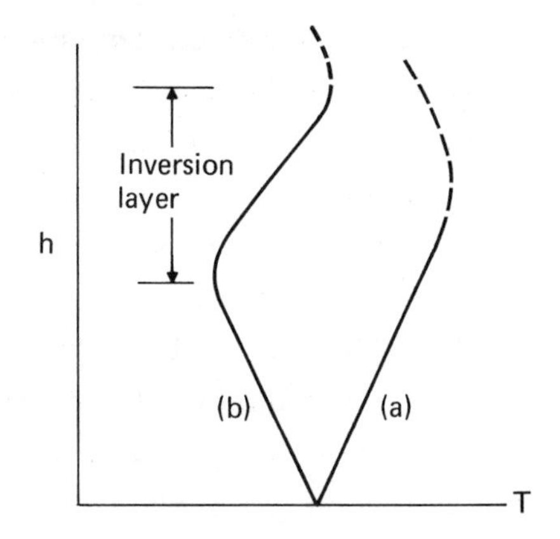


Fig. 3. Inversions

layer. In case (a) the pollutants would be concentrated near the ground, while in case (b) the pollutants often can rise until the inversion layer is reached. At that height vertical mixing ceases and there is a rather abrupt change from dirty to clean air. This "lid" effect can often be seen over large cities if an elevated viewing position is available.

Inversion layers are especially bad when a city is in a valley or surrounded by mountains. Because of the topography, horizontal motion of the air is inhibited. The inversion inhibits the vertical motion, so the pollutants collect. Los Angeles and Donora, Pa., are classic examples of this situation.

What causes a temperature inversion? One way in which an inversion is formed is by radiative cooling on a clear night. The ground is a very effective radiator of heat to space on a cloudless night, whereas air is a poorer radiator. Thus the ground can cool more rapidly than the air. The cool ground then cools off the lower air layers by conduction, leading to the situation of Fig. 3(a). The presence of clouds prevents direct radiation to space by the ground, and so inversions are less likely to form on cloudy nights. This mechanism explains why air quality is often poorer in the morning. An inversion forms during the night due to the ground cooling more rapidly than the air. During the morning hours the sun warms the ground more rapidly than the air, since ground is a good absorber while air is a poor absorber of solar radiation. The air near the ground gets warmed from the ground by conduction, and the inversion is broken up. Often this happens around noon time, so air pollution levels are lower in the afternoon than morning.

Another way in which inversions are formed is through a phenomenon called high pressure subsidence. In general there are regions of higher and lower atmospheric pressures at different places. Air flows out from a "high" towards a "low" (but not in straight lines since the earth is rotating.) This flow occurs mainly at the lower altitudes. As air flows out of a high pressure region near the ground, there is a replenishment from above in the center of the high as shown. As the upper air moves down it is compressed and thus warms up. The subsidence usually does not proceed to ground level, so a region of warmer air is formed some distance above the ground, as shown in Fig. 3(b). Some other causes of inversions are discussed in Ref. 3.

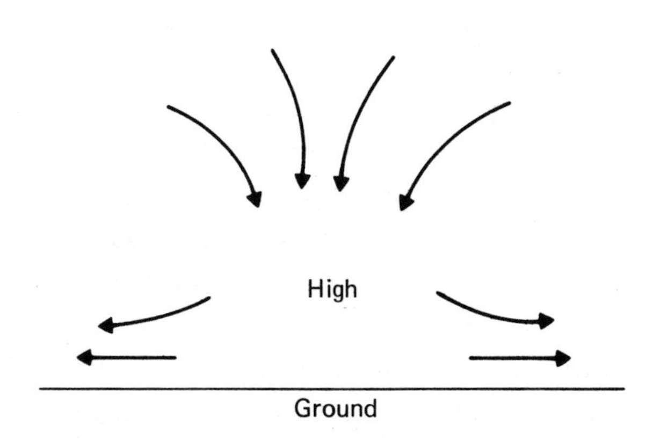


The worst air pollution levels are usually found when a high pressure area is stagnant (not moving). There is not only the subsidence effect mentioned above, but also high pressure areas often mean clear skies which allows the night-time cooling of the ground, and usually there is little or no wind to help disperse the pollutants. With conditions remaining this way for several days, air pollution can reach quite high levels.

High winds help break up a region of stable air, since they are usually accompanied by turbulence. The turbulence will mix the air vertically: some is carried up, some is carried down. In each case the air responds adiabatically so the adiabatic lapse rate is quickly established.

A brief discussion of the stratosphere is now in order. The lower part of the atmosphere, in which the temperature normally drops with altitude, is called the troposphere. Above about 7 km at the poles and about 17 km at the equator, the temperature at first remains constant and then starts rising again. The increase is caused by energy absorbed from sunlight by ozone ($O_3$). This region, which lasts until a height of about 50 km, is called the stratosphere. Since the lapse rate in the stratosphere is zero or negative, this region of the atmosphere is very stable. This stability is one reason why many people are concerned about the SST: pollutant material injected there tends to remain for much longer than is the case for lower altitudes.

REFERENCES

1. J. A. Day and G. F. Sternes, Climate and Weather (Addison-Wesley, 1970) Ch. 6, especially pp. 173-174.

2. W. Bach, Atmospheric Pollution (McGraw-Hill, 1972) Ch. 2.

3. S. J. Williamson, Fundamentals of Air Pollution (Addison-Wesley, 1973) Chs. 5 and 6.

PROBLEMS

1. Consider the case of the lapse rate for saturated wet air quantitatively. You will need information on the saturation of air as a function of temperature.

2. Find the height to which air which is 10 °C warmer than its surroundings will rise if the prevailing atmospheric lapse rate is
   a) adiabatic
   b) 0
   c) -.5 °C/100 m (an inversion).

3. Why are some of the worst air pollution conditions in the winter? For example the following air pollution incidents in which many died were in the cooler months: Meuse Valley, Belguim, Dec. 1930; Donora, Pa., Oct. 1948; London, Dec. 1952; and New York, Nov. 1953.

4. Find an expression for how pressure varies with altitude. If temperature is considered constant (as is often done in textbooks), one can integrate Eqn. 3 to obtain an exponential law. However, as we have seen, the assumption of constant temperature is not very realistic in the troposphere. An alternative result may be obtained if we assume the dry adiabatic lapse rate. This can be done by combining Eqn. 4 with Eqn. 3 or by integrating Eqn. 6 and then using Eqn. 3.

# Unit 14
# Thermal Pollution

## WASTE HEAT

Thermal pollution involves the dumping of unwanted or waste heat into the environment. Electrical power plants are the main source of the waste heat. Since fossil fuels are in limited supply (see Appendix 3), and the demand for electricity is rising, why should any of the valuable energy be wasted?

The answer lies with the second law of thermodynamics. The Kelvin formulation of the second law states that it is impossible to extract heat from a reservoir and convert it entirely into work by any cyclical process. Some of the thermal energy must be rejected to a lower temperature reservoir. The reverse process, converting mechanical or electrical energy entirely into heat, is, of course, possible.

The reversible Carnot cycle allows a simple calculation of the portion of the input thermal energy which can be converted to mechanical work. The work output divided by the heat input is called the efficiency, e. It can easily be shown by direct calculation, or by use of the entropy formulation of the second law, that the efficiency is:

$$e = 1 - T_c / T_h \tag{1}$$

where $T_h$ and $T_c$ are the temperatures on an absolute-zero scale of the hotter and colder reservoirs, respectively. Further use of the second law can show that no heat engine operating between these two fixed temperatures can be more efficient than this. The efficiency approaches 1 as $T_c \to 0$ or $T_h \to \infty$, and approaches 0 as $T_c \to T_h$. The Carnot cycle is not actually used in practice. Most steam engines follow more closely the Rankine cycle. The efficiency of a Rankine cycle is somewhat less than for a Carnot cycle. See any thermodynamics text for more details. We will assume a Carnot cycle for simplicity in this Unit, however, as the conclusions are not altered much.

Clearly there are practical limits to both $T_h$ and $T_c$. Obviously $T_h$ must be kept below the melting temperatures of the materials from which the heat engine is made. If steam is being used in the cycle, as is usually the case now, the limits are even lower. This is a simple result of the fact that steam pressure goes up dramatically as temperature rises. For example, for saturated steam the pressure goes from 14.7 lb/in.$^2$ at 100 °C to 1255 lb/in.$^2$ at 300 °C, and rises to over 3000 lb/in.$^2$ by 370 °C. At the same time the strength of most materials decreases with increasing temperature. The upper temperature can be made higher through the use of some gas other than water vapor and gas turbines.

Presently the temperature limit for gas turbines is about 1200 °C. Another alternative involves the use of the magnetohydrodynamic process (MHD), where temperatures as high as 2000 °C may be possible. See Unit 19. It should be emphasized that these processes merely allow a higher $T_h$; they do not circumvent the second law of thermodynamics. See also Probs. 10 and 11. Current fossil-fuel power plants may have $T_h$ around 500 °C (using superheated steam). Nuclear fission plants operate at lower temperatures, around 300 °C, for safety reasons. $T_c$ is usually near or a bit above the ambient environmental temperature. Problem 3 explores the problems of reducing it below this value.

Power plants do not achieve the efficiences that the foregoing temperatures and Eqn. 1 imply. Equation 1 gives the maximum theoretical conversion. Steam turbines attain only about 80 - 90% of the ideal. In addition there are other losses. Some of the heat goes up the smoke stack. Only between 80% and 90% of the chemical energy in the fossil fuel goes into making steam. There is a small loss in converting mechanical energy from the turbine into electrical energy. We will assume a value of 98% for this efficiency. Thus the overall efficiency for fossil-fuel plants is reduced by the product of the above departures from the ideal case: .8 to .9 for the turbine, .8 to .9 for the stack losses, and .98 for the generator. The overall correction ranges between .63 and .79. We will use .71 as an average correction.

In the case of a nuclear plant, there is no smoke stack. There are some losses in transferring the energy to the turbines, however. These are small, perhaps only 5%, for a boiling water reactor since the heat goes directly into making steam which goes to the turbine. A pressurized water or gas cooled reactor may lose 15% - 20%, since a heat exchanger is used between the reactor and the boiler. The proposed breeder reactors may have two heat exchangers with loses as high as 30%. We will suppose that the turbines obtain 85% of the ideal efficiency and again suppose 98% for the generator. The total departure from ideal thus ranges from .58 to .79. We will use .68 when a definite number is needed in further calculations.

Thus the actual efficiences for an electric power plant are more closely:

$$e = .71(1 - T_c / T_h) \tag{2a}$$

for a fossil-fuel plant and

$$e = .68(1 - T_c / T_h) \tag{2b}$$

for a nuclear power plant. A graph of the overall efficiency as a function of $T_h$ is shown in Fig. 1. A value of 27 °C, corresponding to the ambient temperature on a warm day, was

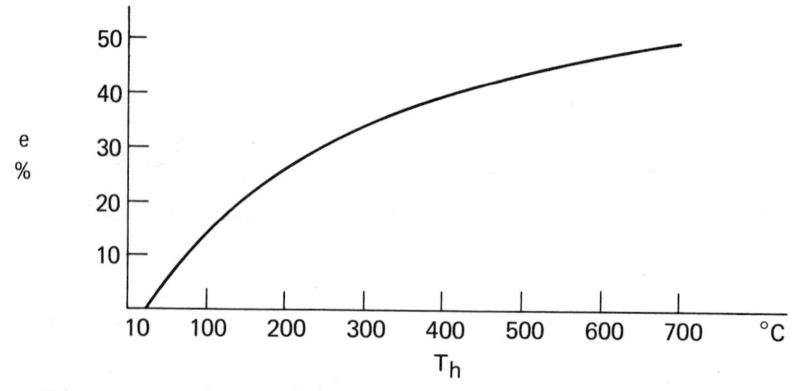


Fig. 1. Overall efficiency of a fossil-fuel power plant as a function of temperature. A nuclear plant is similar.

chosen for $T_c$. The results for a nuclear plant are sufficiently close that a separate curve is not needed.

Two things are apparent from the graph in Fig. 1. Efficiency is below 50% for reasonable temperatures, and as $T_h$ increases, the rate of increase of efficiency decreases. Going from 200 °C to 300 °C increases e by 7.8 percentage points, while going from 600 °C to 700 °C results in an increase of only 2.5 percentage points for fossil fuel plants. So not only is one facing rapidly increasing technological problems as $T_h$ goes up, but the relative gain is decreasing. Modern fossil-fuel power plants obtain about 40% efficiency, while the national average for all plants is about 33%.

While virtually all of the energy consumed in producing electricity eventually ends up as heat (see Prob 4), we are concerned here with the amount of heat rejected by the power plant at $T_c$. Let $Q_c$ stand for the heat rejected at $T_c$, and $Q_h$ for the heat entering the turbine at temperature $T_h$. The work output is:

$$W = .85(1 - T_c / T_h)\, Q_h$$

where we assume 85% for the efficiency of the turbine as compared to the ideal. By energy conservation,

$$\begin{aligned} Q_c &= Q_h - W \\ &= Q_h - .85\,(1 - T_c / T_h)\, Q_h \\ &= (.15 + .85\, T_c / T_h)\, Q_h \qquad (3) \end{aligned}$$

Electric power plants are usually rated in terms of the electrical power output. So we now seek a relation between the heat rejected to the cooling water and the electrical energy output E. We will assume $E = .98\,W$ as before, or:

$$E = .83\ (1 - T_c / T_h)\ Q_h$$

If we eliminate $Q_h$ between the last equation and Eqn. 3, we obtain:

$$Q_c = \frac{.18\ T_h + 1.02\ T_c}{T_h - T_c}\ E \tag{4}$$

This expression is valid for both fossil-fuel and nuclear plants, provided we assume 85% of the ideal thermal efficiency and 98% conversion of mechanical into electrial energy. Some numerical values for Eqn 4 are

$$Q_c = .94\ E$$

for $T_h$ = 733 °K (typical modern fossil-fuel plant) and

$$Q_c = 1.50\ E$$

or 59% more for $T_h$ = **573 °K** (typical nuclear plant). In both cases $T_c$ = 300 °K was used. See also Probs. 7 and 8. A flow chart showing what happens to one unit of energy put into a fossil-fuel plant is shown in Fig. 2 and into a nuclear plant in Fig. 3.

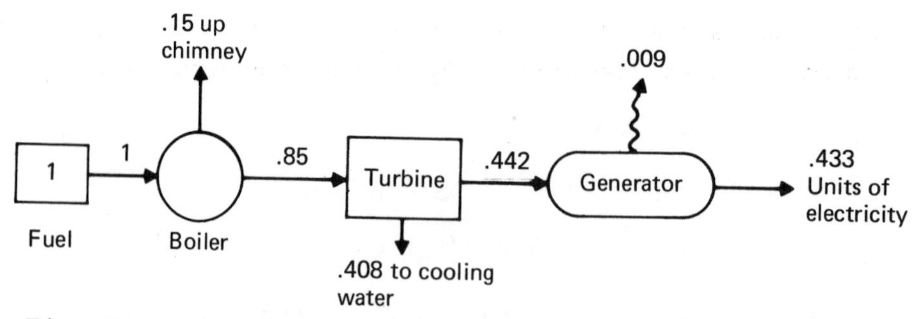


Fig. 2. Diagram of the energy flow in a fossil-fuel power plant. $T_h$ = 500 °C and $T_c$ = 27 °C have been assumed.

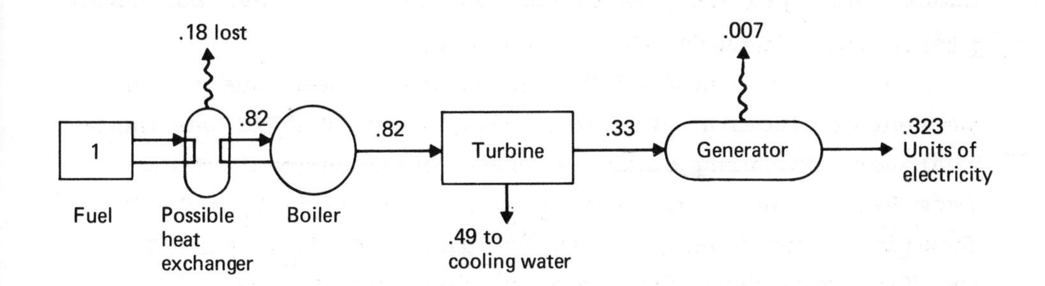


Fig. 3. Diagram of the energy flow in a nuclear power plant. $T_h = 300$ °C and $T_c = 27$ °C have been assumed.

The reader is reminded that we have assumed a Carnot cycle, with approximate corrections added, for simplicity. The actual cases may be somewhat different from our results but are closely similar.

We have not considered the heat wasted in the burning process in Eqn. 4. This goes up the chimney and directly heats the air. Other than adding to the general warming of the microclimates of cities (the "heat island effect"), there seems to be no immediate environmental danger from warming the air. On the other hand the heat rejected at $T_c$ often ends in bodies of water. The warming of streams and lakes may now have significant adverse effects. These are discussed in the next section.

We can take the results of Eqn. 4 to determine how much cooling water a modern 800 Megawatt fossil-fuel power plant needs. The electrical energy output each second is $8 \times 10^8$ Joules. Thus $7.52 \times 10^8$ J or $1.8 \times 10^8$ calories of heat are rejected to the cooling water each second. We will suppose that an 8 °C (14.4 °F) increase in water temperature can be tolerated (exactly how much depends on many factors; see the next section). Then 22500 liters of water are needed each second. This is about 6000 gal/sec or about 800 $ft^3$/sec. Nuclear power plants would require about 60% more.

Before we can find the total waste heat due to our present production of electricity, we must find the waste heat corresponding to the average efficiency of 33%. From Eqn. 2a we find that $e = .33$ implies $T_h = 560$ °K. Equation 4 can then be used to find that $Q_c = 1.56$ E, or 66% more than for a modern fossil-fuel plant.

The present installed capacity for generating electricity is about $4 \times 10^{11}$ W, producing on the average about $1.8 \times 10^{11}$ W. The total amount of cooling water can be found from our $8 \times 10^8$ W example by scaling it up proportionately, and multiplying by 1.66 to take into account the greater waste of the average installation. The result is about $3 \times 10^5$ $ft^3$/sec or $2.6 \times 10^{10}$ $ft^3$/day. This is about 15% of the average daily run-off of water in the U.S., excluding Alaska.

If the consumption of electricity continues to increase at the present rate of 7%/yr (see Appendices 1 and 3), by the year 2000 the average production of electricity will be about $1.1 \times 10^{12}$ W. If we assume that this demand is met 60% by modern fossil-fuel plants and 40% by nuclear plants, and use the figures derived before, we find that about $1.4 \times 10^6$ $ft^3$/sec will be needed. This is about 4.6 times the present need, and amounts to about 72% of the average daily run-off, and may exceed the summertime run-off.

## ENVIRONMENTAL EFFECTS OF WASTE HEAT

In many cases the heat rejected at $T_c$ ends up in some body of water, thus warming it. Since a significant temperature increase will have adverse effects, the term "thermal pollution" is often used for rejected heat. However, not all of the results are bad. We consider here the effects of heat added to water.

The oxygen content of water is critical for most marine life, and it is affected in several ways. First of all the saturation level of dissolved oxygen decreases with temperature. See Fig. 4. At the same time the demands for oxygen increase. If there is dead organic matter in the water, either from natural sources or sewage, oxygen is required for its aerobic decomposition. The amount of oxygen needed is called the biological oxygen demand, or BOD. The rate of decomposition, and hence the rate at which oxygen is used up, increases with temperature. The multiplication rate of plankton increases with temperature, with an attendant demand for oxygen. In general, chemical reactions, including biological ones, increase with temperature. The rate usually doubles with each 10 °C rise in temperature. Fish are cold-blooded and cannot maintain a constant body temperature independent of the water temperature. Thus metabolism rates increase as temperature goes up. The fish are more active and require more food and oxygen. However there is less oxygen available, and the blood is less able to carry it.

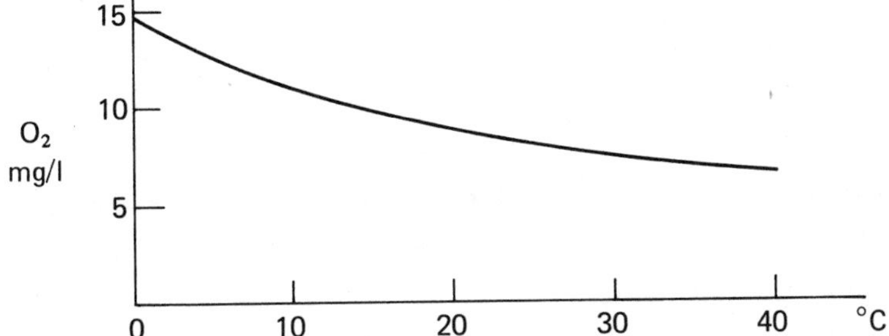


Fig. 4. Maximum amount of dissolved oxygen as a function of temperature.

The toxic effect of various pollutants increases with temperature. For example, a particular fish can tolerate a $CO_2$ level of 120 ppm at 1 °C, but only about 50 ppm at 30 °C. The entire life process is sped up. Eggs hatch faster, fish mature quicker, and do not live so long. This increase in maturation can be used beneficially in fish farms. However, if the temperature is too high, fish will not reproduce at all. At higher temperatures still, the fish die. An upper limit for any type of fish seems to be about 40 °C. Cold water species, like salmon and lake trout die above about 25 °C. Fish can adapt to changes in temperature if it is done slowly enough. Rapid changes, such as would occur if the power plant which had been providing a warm environment shut down for maintenance or equipment failure, can be fatal.

A change in temperature can alter the entire food chain. Algae are a common primary food source. At lower temperatures (below 25 °C) a type of algae called diatoms predominate. As temperature rises, first the greens, then the blue-green algae become dominant. The latter are much less desirable in the food chain. The timing of the food chain can also be thrown off, since maturation rates are altered: a critical food may come too soon or too late in a fish's cycle. Pathogenic organisms can thrive better in warmer waters.

Another effect, which is not thermal in nature, but related to the use of water from rivers to cool power plants, is that flow speeds at the intake and discharge are increased. This can result in the silt and sand being washed away from the bottom, thereby removing the habitat of many forms of marine life.

Because of all of the possible adverse effects of

increases in temperature, a study was made of the effects of a power plant on the Connecticut River. Briefly, it was found that the effects were not so bad as anticipated. In fact the diversity of marine life near the discharge increased instead of the predicted decrease. One explanation for why things were not so bad as expected is that the hotter water tended to stay near the surface; the lower levels of water were not heated. See Ref. 2 for a more complete discussion.

Nevertheless, one can flatly state that if many power plants use water from rivers, there will be undesirable changes in the marine life. The only question is how much is "many"? Fortunately thermal pollution can be avoided; see the next section.

The situation with the use of lake water for cooling power plants is somewhat different. We assume a rather deep lake with slow inflow and outflow of water. The important point concerns the stability and stratification of the water. [The reader should review the idea of vertical stability from Unit 13 if needed.] We start with the fact that water is most dense at 4 °C. Water both warmer and cooler than this is less dense. The lower part of a deep lake thus tends to be 4 °C all year. As the water warms from the top in summer, the warm water is less dense, and remains on top. When the upper layers are below 4 °C in the winter, they again remain on top. Thus a lake has vertical stability and is stratified most of the year. It is only during the spring and fall when the entire lake is about 4 °C that winds can induce large scale vertical mixing. This is the only time when the lower layers can be reoxygenated from the air. During the winter and summer only a negligible amount of oxygen will diffuse down to the lower part. A power plant will draw water from the lower, cooler regions (if permitted), since thermodynamic efficiency is greater, and return the warmer water to the top. The lake will thus become stratified earlier in the spring and remain so later into the fall, thus increasing the time between reoxygenations during the growing season. In extreme cases in moderate climates, the lake could remain stratified year around. While the available oxygen is reduced in the lower levels, the demand increases. This is because the water from the lower levels is rich in nutrients. When brought up to the top, these nutrients increase the marine life in the upper level. When the organisms die, they fall to the lower levels to decompose, consuming oxygen. See also Prob. 6.

More complete details of the environmental impact of dumped heat may be found in the references, especially Refs. 1 and 4.

## CONTROLLING THERMAL POLLUTION

Fortunately, thermal pollution can be satisfactorily controlled with existing technology. The only problem is the cost involved.

Perhaps the simplest method is to use a cooling pond. The warmed water flows into a large pond, where it is cooled by evaporation and contact with the air. After cooling, it may either be reused or flow back into the river. The main problem is the large amount of land needed: for a 800 MW plant about 3 square miles are needed. See Prob. 2. Clearly this is not possible in a heavily populated area, although the pond could be used for recreational purposes. Another problem is water loss due to evaporation and possible seepage.

The main way thermal pollution is controlled is by cooling towers. There are four categories of cooling towers. The main division is into wet and dry, and each of these may either be forced draft or natural draft. In the wet tower, the water is in contact with the air, and cooling primarily takes place by evaporation. We can calculate about how much water is lost by evaporation. The latent heat of vaporization is 540 cal/gm. Cooling 1 gm of water 8 °C requires the removal of 8 cal, which means the evaporation of 0.0148 gm of water, or approximately 1.5% of the water. In our earlier example of a modern 800 MW fossil fuel power plant we found that 800 $ft^3$/sec of water were needed for cooling. Thus about 12 $ft^3$/sec are lost or about one million cubic feet per day if operated at full power. Wet cooling towers do not operate well in a hot, humid climate, because the air is hot and little evaporation can take place since the air is nearly saturated already. A wet cooling tower can cool down only to near the "wet bulb" temperature. Because of all of the moisture placed into the air, fogging, or even local rain or snow can result. One finds signs along some highways reading "Caution Local Fog May Exist Next 1/2 Mile" near cooling towers.

A dry tower is basically a large radiator. The water flows through finned tubes and is cooled by air passing over them. Dry towers are more expensive to construct than wet, but have no water loss.

Mechanical draft towers use large fans to move the air through the cooling tower. The advantage (as compared to natural draft towers) is smaller size and lower initial cost. The disadvantages are higher operating cost and more maintenance. Natural draft towers operate something like a chimney: the warmed air is less dense and rises. Since the humidity has increased, which tends to make air more dense, tall towers are needed. Sizes are on the order of 300 to 400 ft high with proportionate widths.

Costs of the various types of cooling schemes are given in Table 1. The smaller figure is for a fossil-fuel plant and the larger is for a nuclear plant, since there is more waste heat in the latter case. A 800 MW plant is assumed.

Table 1. Cost of various types of cooling. Based on Ref. 3.

| Type | Cost |
|---|---|
| Wet | |
| mechanical draft | $ 7 - 10x10$^6$ |
| natural draft | 9 - 14x10$^6$ |
| Dry | |
| mechanical draft | 16 - 25x10$^6$ |
| natural draft | 32 - 50x10$^6$ |
| Cooling pond | 7 - 10x10$^6$ |
| Once-through cooling | 5 - 7x10$^6$ |

These costs should be compared to plant costs on the order of $100 - 300x10$^6$. The cost of cooling in terms of one's electric bill is estimated to add 1% to 5%.

For perhaps the next century the open ocean can be considered an effective place to dump heat. The main problems are that coastal land suitable for power stations is also sought after for residential and esthetic purposes, and the corrosiveness of the salt water. Confined bays, on the other hand, suffer from thermal pollution similarly to lakes and rivers. One complicating factor is that salt water can hold less dissolved oxygen for a given temperature. However, there is some mixing of water with the ocean due to tides.

An alternative concept of handling the heat rejected at the lower temperature is to utilize it rather than waste it. The main problem here is that the rejected heat is of such a low temperature that it is not very useful. Some possible uses are for fish farming, as already mentioned, heating sewage for more rapid treatment, heating greenhouses, and preheating sea water for desalination. The use of the heat for space heating is limited to winter time in the cooler climates. Even so there are problems with the low temperature, and the large distances to transport the water. The rejected heat would be more useable it it were hotter. This, of course, causes a loss in efficiency. If the waste heat is at 100 °C instead of 27 °C, the efficiency drops to 37% from 43%. We have assumed a fossil-fuel plant with $T_h$ = 500 °C for this example.

REFERENCES

1. J. R. Clark, "Thermal Pollution and Aquatic Life" _Scientific American_, Mar. 1969, p. 19.

2. D. Merriman, "The Calefaction of a River" _Scientific American_, May 1970, p. 42.

3. R. D. Woodson, "Cooling Towers" _Scientific American_, May 1971, p. 70.

4. R. H. Wagner, _Environment and Man_ (W. W. Norton, 1971), Ch. 8.

5. A. Turk et al, Ecology, Pollution, Environment (W.B. Saunder, 1972) Ch. 9.

6. C. M. Summers, "The Conversion of Energy" Scientific American, Sept. 1971, p. 149.

[Note that Refs. 1 and 4 have some incorrect figures on the projected amount of waste heat for the year 2000, and are high on the estimate of the water lost by a wet cooling tower.]

PROBLEMS

1. Compare the total rate of waste heat from the production of electricity in the U.S. to the amount of sun light falling on the U.S. Assume 1400 $W/m^2$ for the sun's power flux (this is the total amount, not the amount absorbed).
   a) Use the present average rate of $1.8 \times 10^{11}$ W
   b) Extrapolate this to the years 2000, 2100, and 2200, assuming 7%/yr growth. (See Appendix 1)

2. Estimate the size of a cooling pond needed for an 800 MW plant by estimating the rate of evaporation and cooling by the air. Compare with the figure given in the text.

3. One way the efficiency of a heat engine can be increased is by decreasing $T_c$ in Eqn. 2. Why don't power companies use refrigeration to reduce $T_c$? Give a quantitative treatment of the overall efficiency including the energy needed for refri-

generation.

4. Estimate what fraction of the electrical power produced does <u>not</u> eventually end up as heat.

5. A hydroelectric plant discharges water at about the same temperature as it takes it in. Is this a violation of the second law of thermodynamics? The efficiency of a hydroelectric plant is about 97%. Reconcile this with Eqn. 1.

6. One of the main drawbacks of using a lake for cooling is the stratification and resulting problems with oxygen in the lower layers. Estimate the amount of energy needed to reoxygenate the lower water by pumping air to the bottom. Assume a lake 100 ft deep and 1 mile by 4 miles, and that the bottom 50 ft are depleted of oxygen. Recall that air is only 21% oxygen, that pressure increases with depth, and use Fiq. 4 for the saturation level. What temperature is most reasonable to use? If this must be done twice a year, what is the average power needed?

7. Equation 4 assumed that 85% of the ideal thermal efficiency was attained by the turbine. Rederive that equation if only 80% of the ideal is achieved.

8. Evaluate Eqn. 4 numerically for serveral other assumed values of $T_h$, and draw a graph of $Q_c/E$ as a function of $T_h$.

9. Calculate the amount of air warmed per day by a 800 MW fossil-fuel power plant using a dry cooling tower. Assume that a 10 °F temperature rise is permitted. Will this contribute significantly to heating the local climate?

10. One way to increase the thermal efficiency is to increase $T_h$, which is permitted by using a gas turbine or MHD "topping cycles" as mentioned previously. The exhaust temperature of these processes is still quite high, so that exhaust is used to heat water in a boiler. Show that ideally a two stage process going from $T_1$ to $T_2$ and then from $T_2$ to $T_3$ gives the same overall efficiency as a single process going from $T_1$ directly to $T_3$.

11. Calculate the ideal efficiencies for gas turbine and MHD cycles, assuming the temperatures given in the text of this Unit.

12. Estimate how long the oceans can serve as a heat dump, if a 2 °C temperature rise can be tolerated. List all of your assumptions. The volume of the oceans can be found in Appendix 4.

# Unit 15
# Space Heating, Heat Pumps, and Air Conditioning

## SPACE HEATING

About 18% of the energy consumed in the U.S. is used for space heating. While this is not the largest segment of the total energy consumption, it is an important one. Most homes are now heated by gas or oil, the fossil fuels which are in shortest supply. Already shortages are felt in these fuels during cold weather. In addition, the automobile also competes for the oil supply. It is estimated that about 80% of the world's oil supply (excluding the problematic shale oil and sand tar oil) will be used by the year 2030. The time by which 80% of the U.S. natural gas supply will be used up has been estimated to be 2015. These figures can be greatly altered by changes in the present energy consumption patterns. Changes in space heating and transportation are particularly important.

It is thus worthwhile to explore ways to conserve

on the fuel used in space heating. Better insulation of buildings is one obvious answer. Estimates show that if all buildings were well insulated, perhaps 1/3 of the energy used for space heating could be saved.

It is quite instructive to calculate heat losses for various parts of a house with various kinds of insulation. The basic heat flow equation is:

$$\frac{dQ}{dt} = \frac{k\ A\Delta T}{d} \tag{1}$$

where $dQ/dT$ is the heat flowing per unit time, $k$ is the thermal conductivity, $A$ is the area of the surface, $d$ is the thickness of the surface, and $\Delta T$ is the temperature difference. Various values of $k$ are given in Table 1, where British units are used since we are more familiar with building dimensions in feet, and furnaces are usually rated in BTU/hr. Since electric heaters are usually rated in Watts, we given the conversion factor: 1 BTU/hr = 0.293 W. Note that the values in Table 1 are based on thicknesses in <u>inches</u> and areas in square <u>feet</u>. These units must then be used in Eqn. 1. The conductivity of air is quite low. This low figure is valid for pure conduction in the absence of convection. In large regions convection usually dominates, and the effective conductivity is much higher. A reasonable case can be made that the function of insulating materials is to provide trapped pockets of air where convection is inhibited, while introducing the least amount of other material which can contribute to the conduction.

For purposes of numerical evaluation, we suppose a medium-sized single story house with ceiling area of 1500 $ft^2$, wall area of 1400 $ft^2$, and window area of 96 $ft^2$. We will suppose a temperature difference of

Table 1. Thermal conductivities, in BTU/hr for one square foot of area, a thickness of one inch, and a temperature difference of one degree Fahrenheit.

| Material | Conductivity |
|---|---|
| air | 0.165 BTU/hr |
| rockwool | 0.26 |
| glasswool | 0.29 |
| various other insulation materials | 0.25 - 0.35 |
| asbestos | 0.55 |
| wood | 0.5 - 1.0 |
| cinder block | 2 - 3 |
| brick | 3 - 6 |
| concrete | 6 - 9 |
| glass | 5 - 7 |
| aluminum | 1450 |

50 °F. Heat losses for other temperature differences scale proportionately. The actual temperature difference across a wall is always less than the difference between the average inside temperature and the outdoor temperature. This is because there is a temperature drop in the air near the wall, and the inside surface of the wall may be a few degrees cooler than the indoor temperature. There may also be a small difference between the temperature at the outer surface of a wall and the actual outdoor temperature. These temperature differences depend on the existance of a "dead air" layer. If there is a significant circulation of air, such as a wind blowing outside, this "dead air" layer will be removed and a greater conduction of heat results.

The rates of heat loss for various materials are given in Table 2. The results given for windows need some discussion. It was assumed that the temperature difference across the glass in the case of the plain windows was the full 50 °F. The "dead air layer" effect mentioned above is particularly important for windows. As an example, measurements showed a temperature difference of 14 °F between the center of a room and a point very close to the inner surface of the window. Most of this difference (8 °F) occurred in the last 1 inch. The outer surface of the glass was 2 °F warmer than the outdoor temperature, with a light wind blowing. The difference between the indoor and the outdoor temperatures was 30 °F during this particular measurement. Thus the temperature gradient across the glass was slightly less than half of the difference between the indoor and outdoor temperatures in this case. The reader is invited to make similar measurements in his own dwelling. Even if the results for windows shown in Table 2 are cut in half, it is still clear that a large fraction of the total heat loss occurs through ordinary windows. Registers and radiators are often placed under windows to compensate for this large loss. One must be careful that the air flow from a register does not disturb the "dead air" layer. If the hot air flows close to the surface of the glass, the heat loss is enhanced. Closing the curtains at night helps keep the "dead air" layer trapped and thus reduces heat losses.

The result given for storm windows is an idealized one, since it assumes that there is no convection in the air between the two pieces of glass and no air leaks. The details of calculating the effective value of the thermal conductivity of storm windows (and other multiple layers) are developed in Probs. 3 and 4. Heat loss to the ground has not been included in Table 2, since the average ground temperature (below frost line) is not so cold, and great thicknesses are involved.

The value of insulation is quite evident from examining Table 2. What may be surprising is the effectiveness of just one inch of insulation, and the tremendous heat loss associated with the small area of windows. Aluminum window frames also contribute significantly to the heat loss, as might be expected from the high value of k in Table 1. See Prob. 6. Table 2 shows that there is little to be gained by going to more than 6 in. of rock wool in the ceiling if the walls are uninsulated. Calculations of this sort show where corrective action is most needed.

The heat lost, and hence the energy which must be supplied, scales directly with the temperature difference: the hotter it is inside, the greater the loss. This is the basis for the recent request for everyone to lower the temperature inside buildings. A 10% reduction in energy used means that the average temperature difference must be reduced by 10%. In the warmer parts of this country, this may mean a reduction of less than 3 °F, while in the coldest parts it implies the inside temperature must be dropped about 10 °F. Clearly a uniform cut-back in supplies of fuel by 10% will not affect everyone equally. Since space heating takes only about 18% of our energy budget, a 10% savings amounts to only an overall improvement of less than 2%. Much greater savings can be achieved through improvements in insulation.

We now consider some of the various ways to heat a building. One can burn fuel directly, use electrical resistance heating, or run a heat pump. Conventional furnaces waste about 20% to 30% of the energy up the chimney. The conversion of electrical energy into heat inside the building is 100% efficient, but one must consider the rather inefficient process of producing

Table 2. Heat loss through various parts of a house.

| | BTU/hr |
|---|---|
| **Roof** | |
| 1" wood | 37500 - 75000 |
| 1" rockwook | 19500 |
| 6" rockwool | 3250 |
| **Walls** | |
| 1" wood | 35000 - 70000 |
| 4" concrete | 105000 - 157500 |
| 4" cinder block | 35000 - 52500 |
| 3" brick | 70000 - 140000 |
| 3" brick and 1" rockwool | 14400 - 16100 |
| 3" brick and 2" rockwool | 8000 - 8500 |
| 1" rockwool | 18200 |
| 3" rockwool | 6066 |
| **Windows (see text)** | |
| 1/8" glass | 192000 - 268800 |
| 1/4" glass | 96000 - 134400 |
| storm windows | 780 |

the electricity, which runs 30 - 40%. Thus over all, resistance heating is about 35% efficient. In spite of the fact that electrical heating is wasteful of energy resources, it has been promoted heavily by the electric power industry as the "modern" way of home heating. It has been estimated that about 18 million homes in the U.S. will have electric heating by 1980, and use as much electricity for heating as the entire home consumption of electricity in 1960. One thing in favor of electric heating is that coal and nuclear fuels and solar energy may be used as the primary energy source instead of the more rapidly vanishing oil and gas.

The use of solar heat for the direct heating of houses

has been suggested for the sunnier winter climates. A rather large expense is involved for an effective solar heating system with overnight energy storage at the present. Even so, back up systems are needed. Usually the need comes during the coldest, stormiest times, when there already is a strain on fossil fuel supplies. See Unit 22 for more details on solar heating.

HEAT PUMPS

Heat pumps have not had much development, but have an interesting theoretical potential. Basically a heat pump is a reversible heat engine run backwards. An amount of work $W$ is put in, heat $Q_c$ is removed from the colder reservoir at $T_c$, and $Q_h$ is put into the hotter reservoir at $T_h$. The Carnot relations apply to an ideal reversible machine:

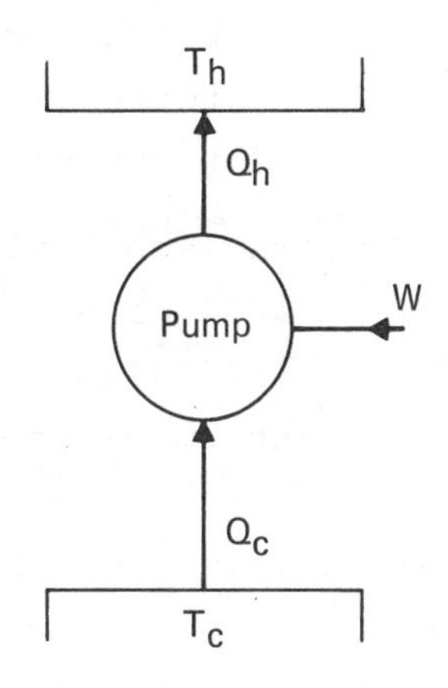


$$\frac{Q_h}{Q_c} = \frac{T_h}{T_c} \qquad (2)$$

Fig. 1. Energy flow for a heat pump.

We are interested here in the relation between $Q_h$ and W. By energy conservation we have

$$W + Q_c = Q_h \qquad (3)$$

Combining this result with Eqn. 2, we obtain:

$$Q_h = \frac{T_h}{T_h - T_c} W \tag{4}$$

If $T_h$ and $T_c$ are close to each other, $Q_h$ can be many times the input work W. For example, suppose $T_c$ = 0 °F = 255 °K and $T_h$ = 68 °F = 293 °K. Then $Q_h$ = 7.7 W. If this ideal could be reached, one would get nearly 8 units of heat out for every unit of energy put in! The most likely source of energy to run the heat pump is electricity. Let us assume 35% efficiency for the generation of electricity. One still obtains 2.7 units of heat for every unit of energy put into the elec-power plant.

Unfortunately, actual heat pumps do not now come close to the ideal efficiency. If they could be developed to half of the ideal, one would get 1.35 units of heat for every unit of primary energy, as compared to .75 units for furnaces and .35 units for resistance heating. If this level of efficiency could be obtained, only 55% as much fuel would be needed by heat pumps as for furnaces and only 26% as for resistance heating. Another major advantage is that heat pumps need not deplete oil and gas supplies. The electrical energy can be generated from relatively abundant coal or by nuclear or solar means. At the present stage of development of heat pumps, some back-up system is still needed, however, because of the low efficiency and limitations on the size of the outside heat exchanger.

## AIR CONDITIONING

Much of the recent increased demand for electricity has come from air conditioning. Indeed, the electrical shortages or "brown outs" occur mostly during the summer (except for those caused by mechanical failure or storms). An air conditioner is basically also a heat pump (or refrigerator), but now we are concerned with the relation between $Q_c$ and W. We use Eqns. 2 and 3 to obtain:

$$Q_c = \frac{T_c}{T_h - T_c} W \tag{5}$$

If we take $T_c = 75\ °F = 297\ °K$, and $T_h = 95\ °F = 308\ °K$, we find $Q_c = 27$ W. Problem 2 explores how closely actual air conditioners approach this ideal.

Air conditioning is often used unnecessarily, for example when the outside air temperature is at or below room temperature and simple ventilating fans would suffice. Modern office-building architecture features many windows, but without proper control of sunlight. Thus rooms on the sunny side are unnecessarily over-heated. The current trend to more lighting, beyond what is required, also adds to the interior heat load, especially when incandescent lights are used. Less than 5% of the energy put into an incandescent bulb comes out as light: the rest is heat. A fluorescent light is still only about 20% efficient.

PROBLEMS

1. An air conditioner was advertised as "22000 BTU capacity". What physical unit is missing? As a clue, the input power was stated to be 2760 Watts.

2. Calculate the ratio of heat removed to input energy for the air conditioner of Prob. 1. You will have to supply the correct missing physical unit before

this problem can be done. Note that the advertised figures do not specify a maximum temperature difference for which the rating is valid. Other air conditioners have ratings which give efficiencies from .72 to 1.26 times the efficiency of the one in this problem. It pays to check ratings!

3. If there are N layers, each of thickness $\ell_i$ and with thermal conductivity $k_i$, prove that the overall effective thermal conductivity is:

$$k_{eff} = \frac{\sum_{i=1}^{N} \ell_i}{\sum_{i=1}^{N} \frac{\ell_i}{k_i}}$$

In using this value of $k_{eff}$ in Eqn. 1, the thickness d is equal to the total thickness:

$$d = \sum_{i=1}^{N} \ell_i$$

4. Calculate the heat loss for storm windows. Assume that there are two layers of 1/8" glass separated by 1" of air. Use the result of Prob. 3. Assume 96 $ft^2$ of area and a temperature difference of 50 °F as before.

5. Find the heat loss through the walls for the house of Table 2 if the walls are 1/2" wood with 2" of glasswool insulation. Use the result of Prob. 3.

6. Aluminum framing is often used for windows. Find the heat loss with the following data (or similar data from your own home):

8 windows

16 linear feet of aluminum frame per window, 1/8" wide and 3/4" thick (the latter is the $d$ for Eqn. 1)

$\Delta T = 50\ ^\circ F$

# Unit 16
# Air Pollution IV: Electrostatic Particle Precipitators

The detrimental effects of particulate matter in the atmosphere and mechanical methods for removing particulates were discussed in Unit 4. In this Unit we consider electrostatic precipitators. These devices are highly effective (up to 99+% efficient) in removing solid and liquid particles ranging in size from about $10^{-3}$ cm to $10^{-5}$ cm. They are not new; the first application to a coal-fired power plant was in 1923 by the Detroit Edison Company. The recent moves to clear up the air have brought them into prominence.

The basic principle is very simple: unlike charges attract each other. Most of the particles in flue gases are neutral, however. Thus one must first place charges on the particles. In usual practice the charging and collection occur in the same region. See Fig. 1. The particles are charged by ions which are created in a corona discharge. Air normally is a rather good insulator with only a few ionized molecules (caused by cosmic rays and background radioactivity). However, if there is a high enough electric field, the few free electrons present can be accelerated to large enough energies between collisions that an electron can ionize the next molecule with which it collides. Thus more and more free electrons are created. If the field is non-uniform, as is the case surrounding a wire, the field strengths at some distance from the wire may become too weak for further production of ions. The ionized region is thus limited in spatial extent. Such a situation is called a corona discharge. (The name comes from the crown-like appearance of the light emitted from the ionized region surrounding a point.) If the electric field strengths are high enough or if the field is uniform, the ionization can extend all the way between the high voltage point and ground and a spark results. The field strengths needed for a corona discharge are estimated in Problem 1, and depend on the air pressure and temperature. A complete discussion of corona discharges would be quite lengthy.

The particles in the dirty air become charged from the ions (we include free electrons as ions for convenience here) present in the tube. Only ions of the _same_ sign as the central electrode are found in the region outside of the corona. This is because ions of the opposite sign are created only in the corona region and travel inwards, while charges of the same sign must go to the outer conductor. Since the discharge occupies only a small portion of the total volume, this means that the dirt particles predominantly become charged with the same sign as the central electrode and are thus mostly attracted to the outer electrode.

One of the two common geometries is shown in Fig. 1,

where a wire and cylinder are used to obtain the necessary non-uniform field. The other common geometry uses wires and flat plates. Typical voltages range from 40 000 to 100 000 V. The weight is used to keep the wire straight without needing another insulator. Can you think of a reason why a lower insulator would be undesirable?

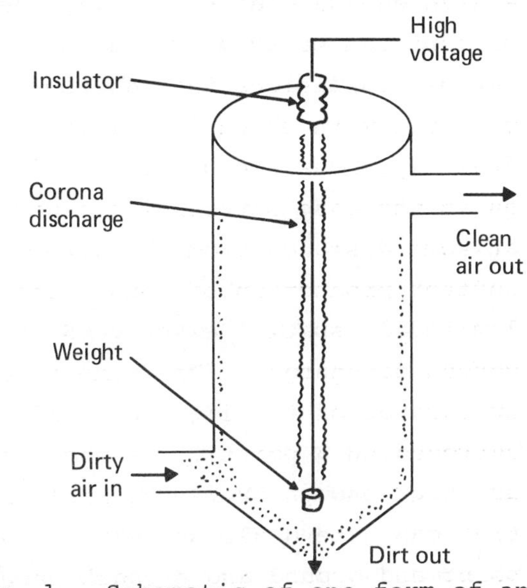


Fig. 1. Schematic of one form of an electrostactic precipitator.

The usual industrial practice is to have the central electrode negative. The negative discharge tends to be more stable and have fewer sparks for a given voltage. For air conditioning dust removal the central electrode is positive since less ozone, which is toxic and causes materials to deteriorate, is formed.

There are some problems with the electrostatic precipitator. The walls obviously get coated with particles which must then be removed by mechanical means such as shaking. There is a problem with the particles' becoming reentrained in the air stream during this process. Often the particles have conglomerated to form larger particles that do not easily enter the air stream, however. Obviously electric power must be used to run the electrostatic precipitator. Typically 50 to 500 W are needed for every 1000 cubic feet per minute processed. Thus more fossil fuels are consumed, more pollution is created, and the costs must be passed on to the consumer.

A rather knotty problem has arisen with electrostatic precipitators. They do not eliminate sulfur oxides, which are the major air pollution problem in many cities. Thus low-sulfur coal and oil have been required by law in many areas. However it has been found that electrostatic precipitators do not work on the fly-ash from low-sulfur coal. To make them work, sulfur oxides are actually added in some cases! Thus it seems that one is caught between having either particles or sulfur oxides. The reason for the failure with low-sulfur coal is not entirely clear, but it is believed that the accumulated particles on the walls have too low a conductivity and thus act as an insulator when sulfur is not present. A large voltage drop in a small region of space results. The field strengths may become large enough that another corona discharge may start, causing the particles to reenter the airstream. Higher temperatures and other conditioners than $SO_3$ may solve this problem. There can also be a problem when the conductivity of the particles is too large (as in the case of carbon particles). Then there may be insufficient electrostatic forces to hold the particles to the collecting electrode and reentrainment may result.

Very hot gases and pressures other than atmospheric require special consideration which will not be covered here.

REFERENCES

1. A. D. Moore. "Electrostatics", Scientific American,

March 1972, p. 47.

2. C. F. Gottschlich. Air Pollution, Vol. III, Ch. 45. (A. C. Stern, editor, Academic Press, Second Edition, 1968).

PROBLEMS

1. Estimate the electric field strength needed to start a corona discharge at standard pressure and temperature. Assume that perhaps 10 eV is the energy needed for ionization and that electrons have about the same mean free path between collisions as gas molecules. The answer obtained in this manner will be somewhat larger than what is actually needed. Why?

2. Find the operating voltage of a concentric-cylinder electrostatic precipitator with the following assumptions. The outer radius is 25 cm, the radius of the inner wire is 2 mm, and the field strength must be at least $3 \times 10^6$ V/m at a distance of 1 cm from the center.

3. Estimate the total power needed to run electrostatic precipitators for a 1000 Megawatt (electric output) power plant. Take into account the overall efficiency of the power plant (see Unit 14), the amount of coal needed to produce the energy, and the amount of air needed to burn that much coal. Then use the average figure of 250 W/1000 $ft^3$/min of air.

# Unit 17
# Transportation VII: Electric Cars

It is generally acknowledged that automobiles contribute the majority of the urban air pollution (by weight). Estimates run as high as 90% in some cities. The source of the pollution is, of course, the present form of the internal combustion engine. One of the alternatives to the internal combustion engine which is frequently mentioned is the electric car. There are many environmentally attractive features: essentially no air pollution produced while running the car, less noise pollution, and less urban waste heat. Other advantages include smooth acceleration, low maintenance costs, and flexibility in the placement of the motor and batteries. In fact, back at the turn of the century, electric cars out-numbered internal combustion ones!

What happened then? Why don't we have electric cars today? The main answer is the limited top speed and driving range of a car based on acceptably priced batteries. For moderate cost batteries the weight needed for a reasonable amount of energy storage is too large. The specific energy of a typical lead-acid battery is about 20 W-hr/lb for a 20 hour discharge time. If the energy is withdrawn more rapidly, the specific energy goes down, so that for powering an automobile 10 W-hr/lb is more typical. See Reference 3.

In Unit 5 we found the specific energy required was ($1.41 \times 10^{-3} vd$) in kcal/lb of vehicle weight. Converting to W-hr/lb from kcal/lb, we have ($1.64 \times 10^{-3} vd$). In these expressions, $v$ is in mph and $d$ in miles. If we assume a small car of 2000 lb and desire an average speed of 40 mph for 100 miles (all of which are modest requirements), we obtain for the energy needed:

$$E = (2000)(1.64 \times 10^{-3})(40)(100) = 1.3 \times 10^{4} \text{ W-hr}$$

This figure is based on steady speeds on the level. In actual driving there are periods of acceleration and hill climbing which require more energy. Some of this energy can possibly be recovered during decelerations and downhill periods. An estimation of the additional energy needed for accelerations, even allowing for partial energy recovery, amounts to about 25% more in town conditions. Thus an energy storage of $1.64 \times 10^{4}$ W-hr is needed. For lead-acid batteries to supply this much energy, 1640 pounds, or 82% of the car's assumed weight, would have to be in batteries! There would be some modest weight-savings in an electric car over an internal combustion car such as less heat and sound insulation, but not enough to compensate. Increasing the overall weight of the car does not change the percentage figure any since the energy required is proportional to the car's weight. The only solutions are to decrease the range and/or average speed or to go to some other energy source than lead-acid batteries.

Some of the other possible batteries under consideration or development are nickel-cadmium, silver-zinc, zinc-air, sodium-sulfur, and lithium-chlorine. Up to about 200 W-hr/lb of battery weight have been reported for the last one. Using the above numerical example for energy requirements, we see that only about 82 pounds of batteries would be needed in that case, which is quite reasonable. A 200 mile trip at 60 mph would require only about 10% of a car's weight to be in batteries. The sodium-sulfur and lithium-chlorine batteries are still under development. The main disadvantage is that they must be mainted at high temperatures (about 250-300°C) since the sodium or lithium must be liquid, which requires energy and would be a hazard in case of an accident. These metals burn when exposed to air or water. Sodium sulfide is toxic. Chlorine is also poisonous. Some of the other batteries mentioned are extremely expensive to make. At the turn of the century it was stated that all the electric automobile needed was a better battery. We are still waiting.

Perhaps the largest factor against the electric automobile is that the total electricity production in the US would have to be increased by a large percentage. See Problem 1. Power plants now are major sources of air and thermal pollution. They and the transmission lines use up land area. A large increase in the number of power plants would exacerbate these problems. Even without the electric car many areas now face shortages of electricity ("brown outs") during the summer. Another environmental problem concerns the fumes emitted by some batteries while they are being charged.

The overall energy conversion efficiency of electric

cars should be examined. First one must convert fossil or nuclear fuel into electricity with an efficiency of .3 to .4. The transmission of the electricity is about 90% efficient. Electric energy must be converted to chemical energy and then back again with about 80%-90% efficiency in each case. The conversion of electrical energy into the finally desired mechanical energy has an efficiency of about .9. The overall efficiency is the product of these numbers, which gives a result of .15 to .26. For comparison an internal combustion car is about 15% efficient. Thus the electric car would use up natural resources at about the same rate or slightly less.

The fuel cell is closely related to the battery and would represent a very attractive alternative if fuel and cost problems can be overcome. Basically one can think of a fuel cell as a battery in which the active chemicals are continuously supplied from the outside and the waste products continuously removed. Efficiences of conversion of chemical energy to electrical can be as high as 75%. High enough specific energies are obtainable. The type of fuel cell which is the most developed uses hydrogen and oxygen or air as the active chemicals. Such cells are used in the space program, for example. The difficulties here are that the present forms are too expensive for automotive use and that there are problems in carrying sufficient quantities of hydrogen in a car. However there is a promising new development which stores hydrogen as a metal hydride, mostly avoiding the temperature, pressure, and weight problems associated with high pressure or cryogenic storage methods. Hydrogen may become readily available as a means of distributing energy; see Unit 21. Fuel cells which use hydrocarbons (e.g., natural gas, gasoline, kerosene, etc.) have not been developed very far. Most actually crack the hydrocarbons first into hydrogen and carbon dioxide. Many of the hydrocarbon fuel cells require high temperatures or expensive catalysts for their operation. Perhaps further development will make hydrocarbon-air fuel cells practical and cheap enough for a satisfactory alternative to the internal combustion engine.

Another approach to electric cars is to carry a generating source with you. Since this would most likely be run by some sort of a fossil fuel engine, it might seem to be a self-defeating idea. However, as was pointed out in Unit 7, with some sort of energy storage it is necessary to meet only the average power requirements, not the peak. An internal combustion engine specially designed and tuned for running at constant speed and load will emit far fewer pollutants than one which must cover a wide variety of conditions. A gas turbine-electric hybrid has some advantages over a straight gas turbine drive. Turbines have low efficiency when not working at high loads. Thus mileage is bad in a turbine car. By having the turbine run a generator and using batteries to store energy, a steady load can be maintained. The main disadvantages of such a system are the high cost and the nitrogen oxides produced by the turbine.

REFERENCES

1. G. A. Hoffman, "The Electric Automobile" Scientific American Oct. 1966, p. 34.

2. Ron Wakefield, "Electric Cars" Road & Track, Jan. 1967, p. 26.

3. "Control Techniques for Carbon Monoxide, Nitrogen Oxide, and Hydrocarbon Emissions for Mobile Sources" NAPCA Publication No. AP-66. Ch. 7.

4. L. G. Austin, "Fuel Cells" Scientific American Oct. 1959, p. 72.

5. T. Aaronson, "The Black Box" Environment, Dec. 1971, p. 10.

6. C. M. Summers, "The Conversion of Energy" Scientific American, Sept. 1971, p. 148.

PROBLEMS

1. Estimate the increase in electric power production required if all vehicles in the US converted to electricity. Assume $10^8$ vehicles and take the present electric production rate at $2.2 \times 10^{11}$ W. Estimate the total amount driven. Use the energy requirement equation of Unit 5. Do not forget the efficiency figures.

2. Following the analysis in Unit 7, estimate the weight of lead-acid batteries needed for temporary storage in an engine-electric hybrid.

# Unit 18
# Build-up of Radioactive Materials

## INTRODUCTION

This Unit employs a powerful technique of physics: the analogy. Frequently there are different areas of physics which involve mathematically identical equations, such as gravity and Coulomb's law, simple harmonic motion and the RLC circuit, etc. Considering these areas as analogies not only provides insight, but saves mathematical labor since the equations need be solved only once. Here we make use of the similarity between RC circuits and the decay of radioactive nuclei. We will simply accept here that radiation is harmful to life (the details are covered in Unit 27). Radiation is always present in the form of cosmic rays and naturally occurring background radiation. Medical uses (x-rays, cancer treatment, etc.) add a significant amount to the natural radiation. The major environment concern is that man may add too much to the unavoidable natural radiation. Atmospheric testing of nuclear weapons threatened to do this before it was halted by the major powers. Now the threat is rising again as man's thirst for more electric power (see Appendix 3) is increasing the number of nuclear fission reactors. Every atom split results in two fission fragments, most of which are radioactive. Disposal remains a problem. Even barring major accidents, some leakage into the biosphere seems inevitable.

## DECAY

In most cases there is little which can be done about radioactive material but to let it decay away. With two minor exceptions (internal conversion and e-capture) no physical or chemical changes can alter the probability of a specific type of nucleus decaying. There is no way that we can tell when a given nucleus is going to decay. All we can give is the probability of decay per unit time. Since we are usually dealing with a large number of nuclei, we can predict the average behavior rather well. The probability of decay is independent of time. This means that a particular nucleus has an equal chance of decaying during the first second of its existence or during its millionth second of existence (assuming that it has not decayed at some time in between).

We can express these ideas mathematically as follows. Let $\lambda$ stand for the probability of one nucleus decaying per unit time and $N$ stand for the number of nuclei of a given type present at time $t$. The number decaying in a time interval $\Delta t$ is the product of the probability per nucleus per unit time, the number of nuclei, and the length of the time interval:

$$\text{Number decaying} = \lambda N \Delta t$$

The number decaying is the decrease or negative change in the number present, so we can write

$$\Delta N = -\lambda N \Delta t$$

On dividing by $\Delta t$ and letting $\Delta t \to 0$ we obtain

$$\frac{dN}{dt} = -\lambda N \quad (1)$$

This differential equation is in the same form as for the charge left on a capacitor in an RC discharge:

$$\frac{dq}{dt} = -\frac{1}{RC} q$$

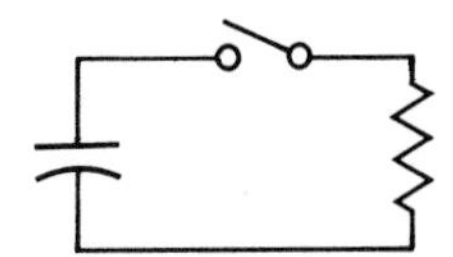

Since the solution to the last equation is

$$q = q_o e^{-t/RC}$$

where $q_o$ is the charge at $t = 0$, we can write the solution to Eqn. 1 by substituting the proper symbols:

$$N = N_o e^{-\lambda t} \quad (2)$$

where $N_o$ is the number of nuclei at $t = 0$. Verifying that Eqn. 2 is a solution of Eqn. 1 is left as a problem. Eqn. 2 gives the correct average behavior. Since this is a probabilistic process there will be random fluctuations about the average behavior.

While one could talk about the $1/e$ time for decay much as is done for RC circuits, it has become more traditional to talk of the half-life, $T_{1/2}$. The half life is the period of time needed for the number of nuclei to reach one half of the previous value. Thus after two half-lives there will be only one half of one half or one quarter as many as originally. In general after n half-lives there will be $(1/2)^n$ as many remaining. See Fig. 1. We can relate $T_{1/2}$ to $\lambda$ by the following. From Eqn. 2 we have :

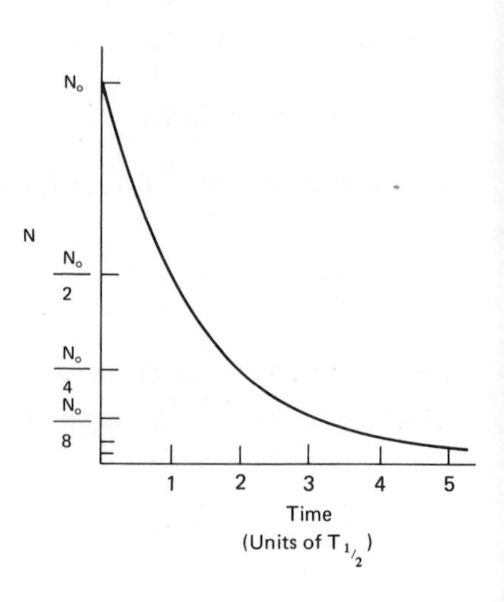


Fig. 1. Number of radioact
nuclei remaining as a funct
of time.

$$N = 1/2\ N_o = N_o e^{-\lambda T_{1/2}}$$

or

$$e^{+\lambda T_{1/2}} = 2$$

Thus

$$\ln(e^{\lambda T_{1/2}}) = \lambda T_{1/2} = \ln(2) = .693$$

so

$$T_{1/2} = \frac{.693}{\lambda} \tag{3}$$

Half-lives range from less than $10^{-16}$ sec to about 10 billion years.

BUILD-UP

We now consider what happens when there are sources of the radio-active material. The environmentally important factor is the decay rate, which is proportional to the number of radioactive nuclei present at any given time. Thus we must find expressions for the number of nuclei as functions of time for various assumed sources. We consider first the case where there is a steady (time independent) source of the radioactive nuclei. This is the simplest case and would be appropriate if there were a constant number of nuclear reactors, for example. In the case where sources are present, the number of nuclei changes with time not only because of decay as in Eqn. 1, but also due to the source. If we let $S$ be the number produced per unit time, then

$$\frac{dN}{dt} = -\lambda N + S \tag{4}$$

This equation is of the same for as for the charging of a capicitor by a battery through a resistor:

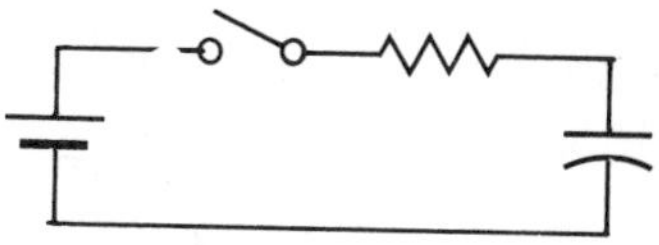

$$\frac{dq}{dt} = -\frac{q}{RC} + \frac{V}{R}$$

Since the solution to the above equation, for $q = 0$ when $t = 0$, is

$$q = CV(1 - e^{-t/RC})$$

we have by analogy for the solution of Eqn. 4 under similar boundary conditions:

$$N = \frac{S}{\lambda}(1 - e^{-\lambda t}) \tag{5}$$

The decay rate is just $\lambda N$, or

$$\text{decay rate} = S(1 - e^{-\lambda t})$$
$$= S(1 - e^{-.693\ t/T_{1/2}}) \qquad (6)$$

The assymptotic value as $t \to \infty$ is

$$\text{decay rate} = S \qquad (7)$$

This result shows that a steady-state condition exists in which the decay rate is equal to the source rate. The decay rate is within 97% of the steady state value after 5 half-lives. See Prob. 5. The behavior of Eqn. 6 for varying S and varying $T_{1/2}$ is shown in Fig. 2 and Fig. 3 respectively. Note that the time for approaching steady state depends only on $T_{1/2}$ and not on S, and that the steady state level is independent of $T_{1/2}$.

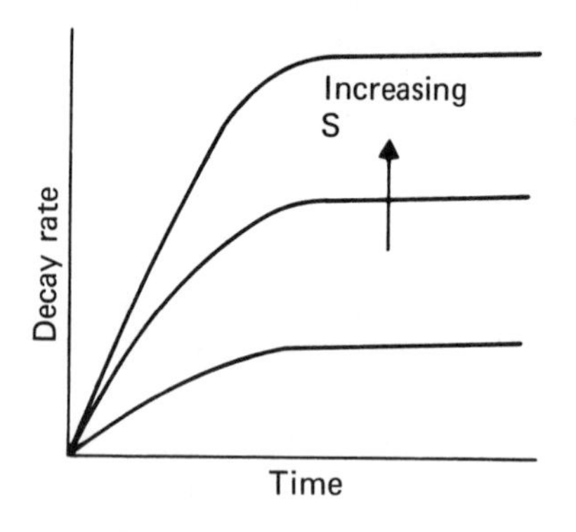


Fig. 2. Effect of different values of S on the decay rate for fixed $T_{1/2}$.

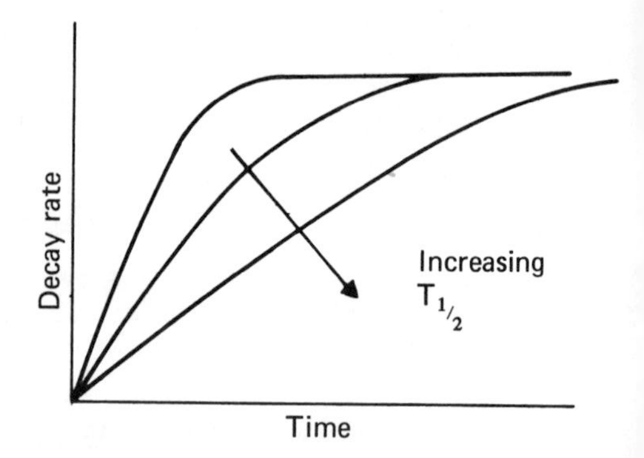


Fig. 3. Effect of different values of $T_{1/2}$ on the decay rate for fixed S.

The radioactive nuclei of the greatest immediate concern are those of moderate half-lives (from a few months to a few tens of years). The shorter lived ones can probably be successfully contained until they have decayed to a harmless level, while the longer lived ones will not build up to their steady state levels for a while. In the long run the radioactive nulcei with large half-lives are the real problem, since they must be confined for a long time.

However alarming the foregoing discussion may be, it unrealistically underestimates the problem. A steady source function is not realistic since the number of nuclear power plants is increasing with time. We can start from Eqn. 4 and assume that the source rate is increasing linearly with time to obtain

$$\frac{dN}{dt} = -\lambda N + at \qquad (8)$$

or we can assume that the source rate is increasing exponentially with time.

$$\frac{dN}{dt} = -\lambda N + c\, e^{bt} \qquad (9)$$

The last equation will probably be the best model for a while, since the consumption of power is closely following exponential growth (see Appendix 3). Of course exponential growth can not go on forever. The solutions to the above equations are: for Eqn. 8.

$$N = \frac{a}{\lambda^2}(e^{-\lambda t} - 1) + \frac{a}{\lambda}t \qquad (10)$$

and for Eqn. 9

$$N = \frac{c}{\lambda+b}(e^{bt} - e^{-\lambda t}) \qquad (11)$$

In both cases it has been assumed that $N = 0$ when

$t = 0$. The verification that these are solutions is left as a problem. The corresponding decay rates are

$$\text{decay rate} = \frac{a}{\lambda}\left(e^{-\lambda t} - 1\right) + at \tag{12}$$

decay

for the linear source, and

$$\text{decay rate} = \frac{\lambda c}{\lambda + b}\left(e^{bt} - e^{-\lambda t}\right) \tag{13}$$

for the exponentially increasing source. The behavior of Eqns. 12 and 13 is plotted in Fig. 4. Note that in neither case is there any steady state condition. The amount of radiation keeps on increasing as long as the sources grow.

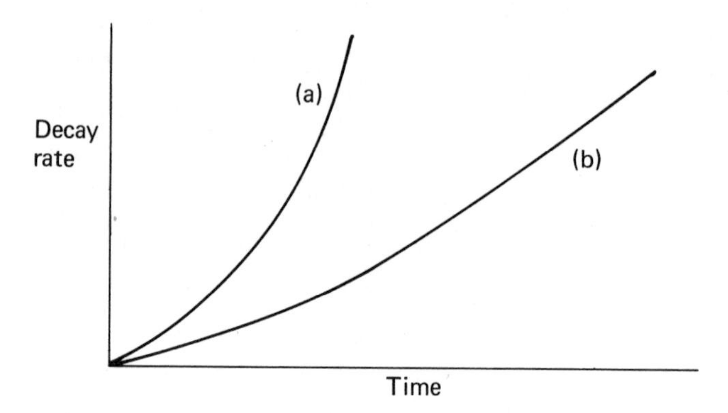


Fig. 4. Decay rates as functions of time for (a) exponentially increasing and (b) linearly increasing sources.

## CONFINEMENT

Our discussion so far has been on the build-up of the total amount of radioactive material and the corresponding decay rates. What fraction, if any, of this material gets into the biosphere is another question. We can only give a brief outline here of some of the considerations.

The inert gases are hard to confine, since they do not chemically combine and are gaseous. They can escape through the unavoidable cracks in the cladding material surrounding the fuel elements in a fission reactor. Storage is a problem even when the gases are caught, since large volumes and/or high pressures are needed for gases, or very low temperatures must be used if liquified.

Many of the other radioactive atoms are readily incorporated into living matter. Tritium (hydrogen-3), carbon-14, strontium-90, cesium-137, and several isotopes of iodine are among the most biologically dangerous ones. The level of activity can be increased by concentration in the food chain. Certain atoms are needed for plant life, and these are gathered from the environment and concentrated in plants. A grazing animal will eat many of these plants from a wide area and concentrate the material further. This process goes one step further when man eats meat.

It appears potentially possible to confine all radioactive material and keep it from the biosphere during normal operation of a reactor. The amount of radiation from the radioactive material confined within a reactor is essentially zero at distances more than 2000 ft from the reactor because of shielding, including the air, and the inverse square law. But accidents are unavoidable, due to human and material failure. The amount of radioactive material which can be released in a large accident is huge. For example about $10^{12}$ times as much strontium-90 is produced in a large power reactor in one year as is permitted in the human body as a safe level. If an accident permitted only one-tenth of one percent of this to escape, and only one-tenth of one percent of that found its way to a large city, there would still be enough to give one million people the maximum safe dose. Note that most power reactors are located close to large population areas since that is where the power is used. It would be easy to imagine worse accidents than the case given above. Other places where accidents could occur are during the transportation of the used fuel elements to the processing plants, in the processing plants, and in the storage areas.

Storage gives many difficulties. The radioactivity creates heat, which sometimes is intense enough to require refrigeration. Much of the material is stored in liquid form, some of which is corrosive. Radiation causes damage to the materials of the storage tank. A storage tank usually lasts only about 20 years. The possibility of earthquakes can not be ruled out in any area. Keep in mind that we are talking about storage which in some cases must last for millions of years. Man has experience with the durability of materials over only very much shorter times. The use of salt mines and the Antartic ice cap for storage has been proposed. There are possible problems with both. See the References.

REFERENCES

1. J. Harte and R. H. Socolow, Patient Earth (Holt, Rinehart, Winston, 1971) Ch. 18, and pp. 132-134.

2. Sheldon Novick, The Careless Atom (Dell Publishing, 1969).

3. R. S. Lewis, "The Radioactive Salt Mine," Bull. Atomic Scientists, June 1971, p. 27.

4. E. J. Zeller *et al*, "Putting Radioactive Wastes on Ice," *Bull. Atomic Scientists*, Jan. 1973, p. 4.

PROBLEMS

1. Verify by differentiation that Eqn. 2 is the solution of Eqn. 1. Similarly for Eqns. 5 and 4, 10 and 8, and 11 and 9.

2. Suppose that the steady source rate for a particular radioactive nucleus is $5\times10^{18}$ nuclei/sec, and that the half-life is 30 years. Note: these figures are about right for cesium-137 from a single large reactor.

   (a) How many radioactive nuclei have been built up after one year? How many disintegrations per second result from this?

   (b) What is the final steady state number of radioactive nuclei?

3. Analyze the asymptotic behavior (that is for $t$ much larger than $T_{1/2}$) of Eqns. 12 and 13. Use some reasonable values to evaluate these asymptotic solutions numerically.

4. If $10^{26}$ nuclei have been produced after 3 years, what are the source rate and the final steady state number of nuclei? Assume a constant source and a half-life of 4 years.

5. Show that after $n$ half-lives the decay rate is within $(1 - 2^{-n})$ of the steady state value for a constant source.

6. How long must strontium-90 be stored if the activity is to go down by a factor of $1.5 \times 10^{12}$? The half-life is 28 years.

# Unit 19
# Electricity Production Using Magnetohydrodynamics

Magnetohydrodynamics is basically the study of the flow of a conducting fluid in a magnetic field. For obvious reasons the term is almost always abreviated as MHD.

Most of the generation of electricity involves a conversion of thermal to mechanical energy. Hydroelectric power is the only significant form of generation that avoids this step at present. Any conversion of thermal energy is limited to the Carnot efficiency or less (see Unit 14).

$$e = 1 - \frac{T_C}{T_H} \quad (1)$$

where the temperatures are in °K, and $T_H$ is the higher temperature. High efficiences are environmentally desirable for two reasons. The larger e, the more the heat serves useful purposes and the less of the heat is rejected to the environment as thermal pollution. Secondly higher efficiences mean less natural resources need be used up for a given amount of electrical energy produced.

Since $T_c$ can not be any less than the ambient temperatures for practical purposes, the only way to increase the efficiency is to increase $T_H$. At present the steam turbine is the main device used for converting thermal energy into mechanical energy. There is a practical limit to how hot steam can be made, since as temperature goes up, so does pressure, and at the same time the strength of most materials decreases. Technological improvements allow the temperature of a steam cycle to be raised only very slowly. What is needed is some way to increase effective temperatures by a large amount.

Enter MHD. MHD is not a new source of energy nor does it circumvent the second law of thermo-dynamics; instead it allows a much higher $T_H$. The principle is as follows. Suppose that a highly ionized gas is flowing into a region where there is a magnetic field perpendicular to the direction of flow. See Fig. 1. [A highly ionized gas will consist of positive and negative ions and electrons, but be neutral overall, and is called a plasma.] There is a force acting on each charged particle given by

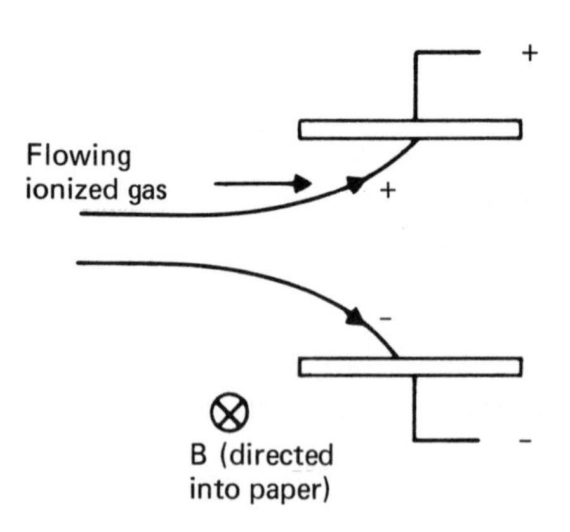


Fig. 1. Basics of the production of electricity by MHD.

$$\vec{F} = q \vec{v} \times \vec{B} \tag{2}$$

The direction of the force depends on the sign of the particle. Positive particles go one way and negative particles the other. The two collector plates shown in Fig. 1 thus become charged up, creating an electrical potential difference. If the plates are connected to an external load, current will flow and energy will be delivered. Where does this energy come from? Since like charges repel, as the plates charge up work is done against the charged particles, slowing them down. Thus the energy comes from the kinetic energy of the particles. The electrical output is obviously direct current. An alternator can easily convert this to a.c. The magnetic field can be supplied by superconducting magnets. Since the magnetic field does no work, little energy need be expended here.

A hot flowing gas can be obtained from a combustion chamber. But at the usual temperatures encountered in combustion (1000 °C to 2000°C), the degree of ionization is very low. Thus a seed material such as cesium or potassium or one of their compounds must be added to the combustion chamber. These elements are easily ionized thermally and can lead to a satisfactory degree of ionization for the gas. Even so, preheated air or oxygen injection appears necessary to achieve high enough temperatures (in the range of 2000 °C to 2500 °C). The rest of the gas becomes partially ionized by collisions with the ionized seed material. The seed material must be recovered before the gases are ducted to the exhaust stack. This is not only because of possible environmental impact, but because the seed material is expensive. At the same time air pollutants such as particulates and sulfur oxides, can be removed. It is expected that the sulfur will react with the seed material to form easily-removed compounds. Some proponents of MHD actually feel that this is the chief advantage of this process, since it allows the plentiful high-sulfur coal to be used without adverse air pollution. Nitrogen oxides are readily formed at the high temperatures involved. It is felt that they can be controlled by properly regulating the combustion process.

Not all of the energy will be converted, of course. Not only is there the Carnot limit, but not all of the gas will be ionized. The waste heat from the MHD convertor can be inputted to a conventional low temperature steam generator, increasing the overall efficiency some. Thus a schematic diagram for an MHD-conversion plant might be as showin in Fig. 2. It is also possible that gas turbines could be used after the MHD cycle, thus avoiding

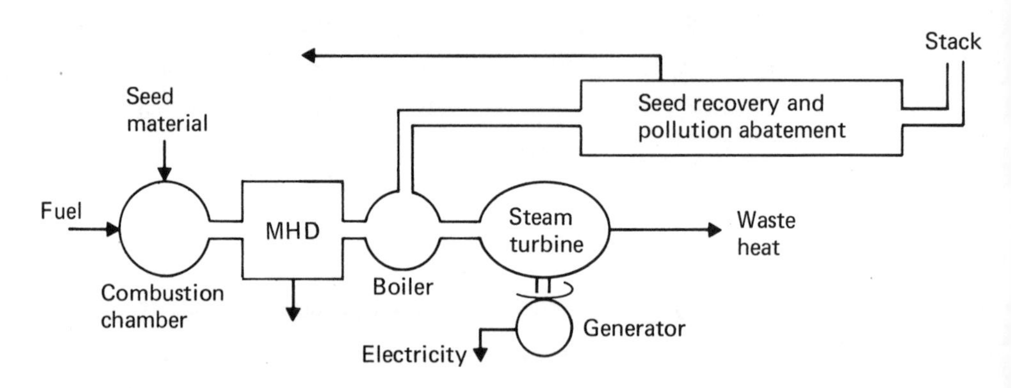


Fig 2. Schematic of an electric power plant using MHD and a steam cycle.

We can get an idea of the conversion efficiency from Eqn. 1. Let us assume that $T_C = 300\ °K$ and $T_H = 2500\ °K$. The ideal efficiency then is 88%. This figure is rather higher than what can be achieved. Various estimates of the overall efficiency of an MHD plant (including the follow-up steam stage) range between 50% and 60%. If the latter figure can be achieved, it would be a significant and worthwhile improvement over the present 30% to 40% conversions. Fossil fuel reserves would last longer.

There does not seem to be any scheme at present which will allow the use of MHD with a nuclear fission reactor. This is unfortunate since we seem to be heading towards an era of fission reactors. However, if controlled nuclear fusion is ever achieved (see Units 20 and 28), the way is again open for MHD. In fact MHD may be the only practical way of recoverying the energy carried by the charged reaction products. [Unfortunately a large portion of the energy goes to uncharged neutrons.] A fusion reactor operates at extremely high temperatures (on the order of tens of millions of degrees) where the gases are completely ionized. Large magnetic fields will probably already be present (see Unit 20). Thus all of the needed ingredients are present. Details must await the development of the successful confinement configuration.

REFERENCES

1. C.M. Summers, "The Conversion of Energy". Scientific American Sept. 1971, p. 148.

2. A. L. Hammond, "Magnetohydrodynamic Power: More Efficient Use of Coal." Science 178, 386 (Oct. 27, 1972).

PROBLEMS

1. What determines the maximum potential difference between the two collector plates? Estimate a

numerical value.

2. What determines the maximum current which may flow from the collector plates? Write down an expression involving temperature, gas flow rates, degree of ionization, densities, or any other parameters needed. Hints: what is the microscopic expression for current? How does average speed relate to temperature?

# Unit 20
# Plasma Confinement for Controlled Thermo-nuclear Reactions

## INTRODUCTION

An overview of man's energy consumption and the resources available is given in Appendix 3. Briefly, the only major energy sources capable of lasting for millions of years are solar energy (Unit 22) and energy from nuclear fusion reactions. The nuclear physics, energy release and resources for fusion are covered in Unit 28. Very high temperatures are needed for fusion reactions, as explained below. At these high temperatures matter is in the form of an ionized gas, referred to as a "plasma". This Unit will consider the problem of confining a plasma long enough for useful power production to take place. Success in controlling fusion will mean abundant power with little pollution associated with it. Unfortunately the feasibility of controlled fusion has not been demonstrated yet. Even if a positive net energy return from a controlled fusion process in the laboratory were to be achieved in the next few years, there would be many problems in scaling up to commerical sizes. It is reasonable to suppose that at least another 50 years would elapse before a significant amount of power would be available and another 30 years beyond that before the majority of our energy could come from fusion. Fortunately other sources of energy can supply our needs meanwhile. However, a major effort must be started now to achieve such a schedule.

Basically fusion involves bringing two light nuclei (usually heavy isotopes of hydrogen) close enough to each other that nuclear forces pull them together into a heavier nucleus. Energy is released in the process. The problem is that the nuclei are positively charged and repel each other. Therefore the nuclei must approach each other with considerable energy. One way of achieving this is to use a nuclear accelerator to impart a high speed to one nucleus. However, the energy used by the accelerator far exceeds the energy released.

The only reasonable alternative is to heat the desired atoms to a very high temperature. The average kinetic energy of the ions is proportional to the absolute temperature. If we express energy in electron volts (eV) and $T$ in °K, the relation becomes

$$\bar{E}_k = 1.29\times10^{-4}T \tag{1}$$

We can estimate the temperature needed as follows. In order for the reaction to have a reasonable chance of happening, kinetic energies of the order of $5\times10^4$ eV are needed (see Prob. 1). It is not necessary that the average kinetic energy be this high, since some of the atoms will have energies several times the average. We thus suppose that the average kinetic energy should be around one quarter of that needed for the reaction. Thus one needs temperatures in the range

$$T = \frac{1}{4}(5 \times 10^{4})/1.29 \times 10^{-4} = 10^{8}\ ^{\circ}K$$

More detailed calculations show that a somewhat lower temperature will suffice if deuterium ($H^2$) and tritium ($H^3$) are used as a fuel, and somewhat higher temperatures are needed if deuterium only is used. See Unit 28 for more details on the fuels.

Achieving such a high temperature, difficult as it may be, is only a necessary but not sufficient condition. Hot gases (and plasmas) like to expand. The plasma must be held together, or confined, long enough for a sufficient number of productive collisions to take place in order for more energy to be released than was consumed. Most of the collisions will not lead to energy producing reactions. In the first place only some of the atoms have enough energy. Even for those with enough energy, there is a large probability that the nuclei will merely have an elastic collision. The question of calculating the probabilities of the various sorts of collisions is beyond the scope of this book. The collision rate depends on the density and the mean speed, increasing with both. J. D. Lawson calculated that more energy will be released than consumed when the product of the density and the confinement time exceeds about $10^{14}$ sec/cm$^3$. This figure is for the deuterium-tritium reaction; it is about ten times higher for the deuterium-deuterium reaction. Two basic approaches are being tried. One is to use a density around $10^{14}$/cm$^3$ (a modest vacuum) and confinement times around one second. Clearly something must be used to keep the plasma from expanding during this time. Ordinary material walls can not be used. This is not because the plasma would melt the walls (the energy content is too small), but because the walls would cool the plasma too quickly. The low density approach uses magnetic fields for confinement.

The other approach is to use high densities, similar to those in solids and liquids, in which case the necessary confinement time becomes so short that the plasma does not have time to expand. This approach is sometimes called inertial confinement. We now consider the two basic approaches in greater detail. Still other alternatives have been suggested. See Reference 8.

## MAGNETIC CONFINEMENT

Most of the effort that is going into fusion research uses the concept of magnetic confinement. As we saw above, at fusion temperatures the fuel is ionized and forms a plasma. Charged particles moving in a magnetic field are acted on by a force

$$\vec{F} = q\,\vec{v} \times \vec{B} \tag{2}$$

We can write the magnitude of the force as

$$F = q\,v_{\perp}\,B \tag{3}$$

where $v_{\perp}$ means the component of the velocity perpendicular to the magnetic field.

Consider a positive particle moving in a plane perpendicular to a uniform magnetic field. The direction of the force is perpendicular to both $\vec{v}$ and $\vec{B}$. Thus no work is done by the force, and by the work-energy theorem, the magnitude of $v$ remains

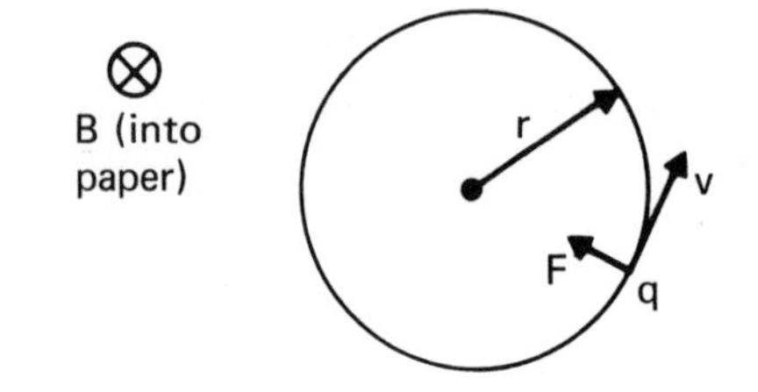


Fig. 1. Motion of a positive particle in a uniform magnetic field.

constant; only the direction changes. So a curved path results. Since by Eqn. 3 the magnitude of the force is constant, the path will be a circle of constant radius. We can determine the radius by equating Eqn. 3 to mass times centripetal acceleration:

$$q\, v_{\perp}\, B = m\, v_{\perp}^{2}/r$$

with the result

$$r = \frac{mv_{\perp}}{qB} \tag{4}$$

If the magnitude of B varies we can still use Eqn. 4 qualitatively. Where B is larger, the radius will be smaller. A motion as shown in Fig. 2 results. There is a general drift perpendicular to the direction of the gradient of B. The reader should verify that a negative particle will drift in the opposite direction. This result will be needed later.

If the velocity of a particle is not perpendicular to the direction of the magnetic field (which will usually be the case), the projection of the motion on the plane perpendicular to the magnetic field will still be as above. The motion parallel to $\vec{B}$ will be unaccelerated. Thus a spiraling or helical motion will result. The foregoing discussion leads to the general magnetic confinement principle. A magnetic field inhibits the spreading of a plasma in directions perpendicular to the field lines. The particles spiral about the field lines.

Motion along the lines proceeds freely, however. There are several ways to "plug the ends." One is to make the reaction chamber so long that end losses are negligible. Estimates of the length needed, however, run on the order of miles. Another solution is to greatly increase the magnitude of the field at the ends. The shape of the field is then about as shown in Fig. 3. This configuration is known as a magnetic mirror or magnetic bottle. In the middle the force on a charged particle points radially inward. At the ends the force, which is always perpendicular to the field, has a component directed away from the ends, which causes the particles to reverse direction. This effect is shown in Fig. 4. The name "mirror" comes about because the particles are "reflected" at the ends. There will still be some losses out the end for those particles with velocities directed very close to the field direction. It has been suggested that the particles leaking out the end could go into an MHD convertor.

The most obvious way to avoid end losses is to have no ends! This is accomplished by bending the chamber into a toroidal or

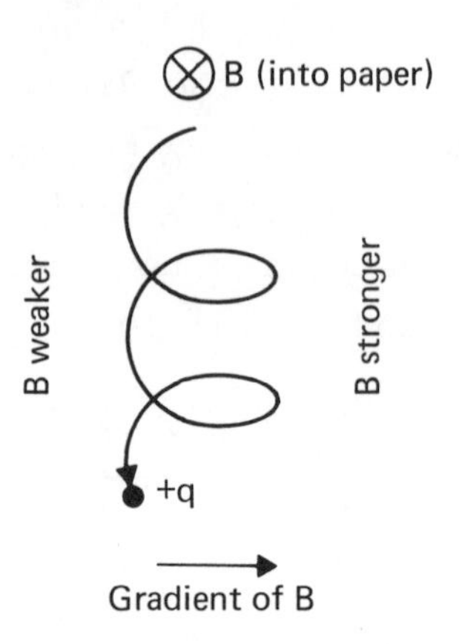


Fig. 2. The drift of a positive particle in a non-uniform magnetic field.

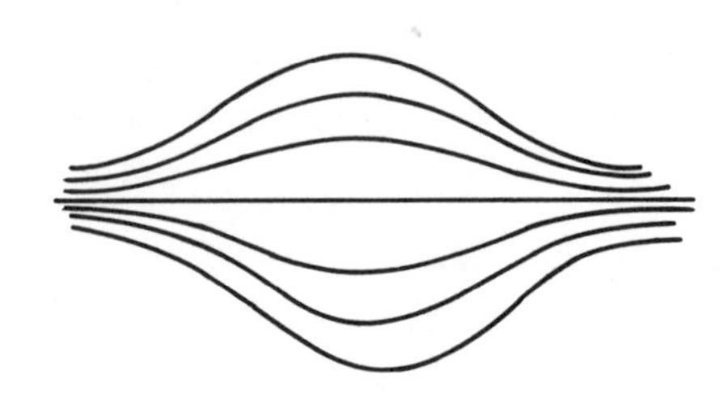

Fig. 3. The magnetic bottle or mirror field shape.

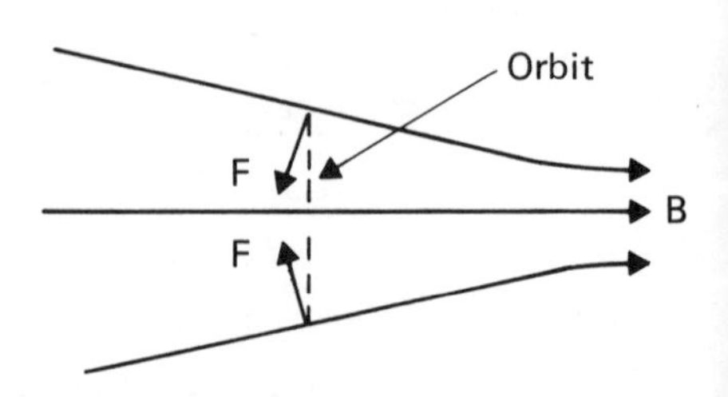


Fig. 4. Details at the end of the bottle.

donut shape. See Fig. 5. A new problem immediately arises. The magnetic field inside a toroid is not constant in magnitude but decreases with increasing radius, as a simple application of Ampère's law will show. Ampère's law is

$$\oint \vec{B} \cdot d\vec{l} = \mu_o i_{tot}$$

The total current is the current per turn times the number of turns. Let the contour of integration be the dashed line shown in Fig. 5. The equation becomes

$$2\pi r B = \mu_o ni$$

Thus

$$B = \frac{\mu_o ni}{2\pi r}$$

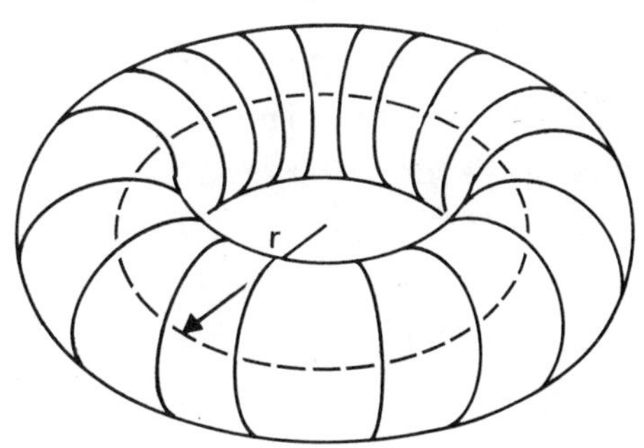


Fig. 5. A toroid.

demonstrating the decrease in B with increase in radius. This gradient in B will cause positive particles to drift one way (say to the top in Fig. 5) and negative particles the other. The results are disasterous to confinement. If an additional magnetic field is supplied so that particles spend part of the time in the higher field and part of the time in the lower, this problem can be avoided. This may be accomplished by external coils as in the Stellarator design (see Reference 11), or by an internal current in the plasma as in the Tokamak design. The latter holds the greatest hope for adequate magnetic confinement for fusion at the present. The current in the plasma is induced by having the ring of plasma serve as the secondary winding of a transformer. The arrangement is shown schematically in Fig. 6. As the current in the primary is varied, the magnetic flux through the iron core, and hence through the center of the plasma ring, varies. By Faraday's law, an emf then appears in the ring causing a current to flow. This current may reach as high as $6 \times 10^5$ A in some of the new designs.

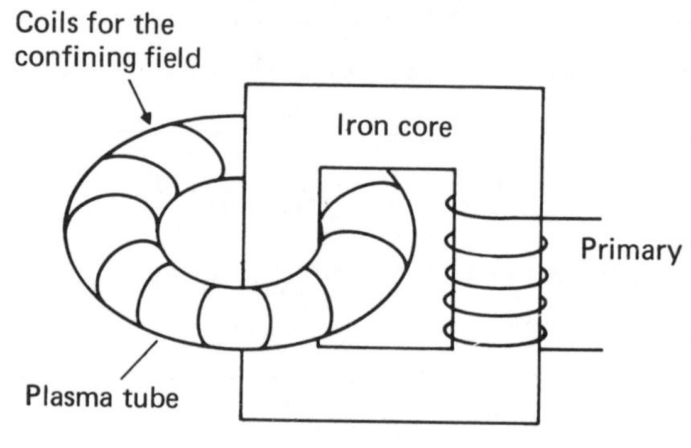


Fig. 6. Schematic of the Tokamak design.

A current flowing in the plasma leads to a general contriction or radial compression of the plasma. This is a simple result similar to the fact that two wires carrying current in the same direction attract each other. Review that section of your text. This effect is called a "pinch".

## INSTABILITIES

The main difficulty with confining plasmas by magnetic fields is the presence of instabilities which lead to a rapid dispersal of the plasma. The class of macroscopic instabilities is reasonably well understood now, and is the simpler to illustrate. The macroscopic instabilites involve the plasma as a whole, considered as a conducting fluid.

An analogy to water shows the general behavior of one class of instabilities, called the "exchange instability." Consider a container with water above and gas under pressure below,

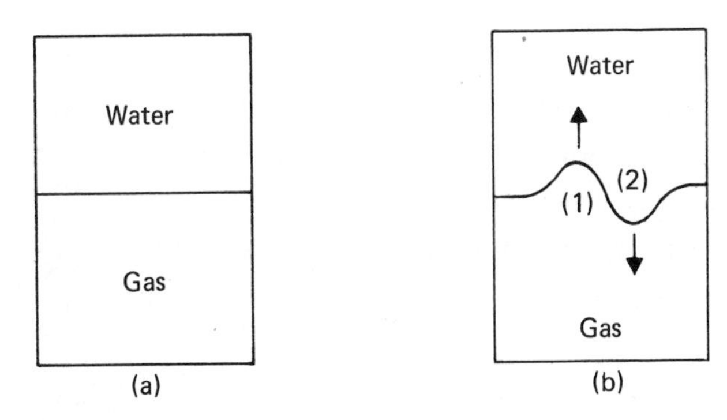


Fig. 7. Illustration of the exchange instability.

as in Fig. 7a. If the lower surface of the water is level and the gas pressure is equal to the hydrostatic pressure at the bottom of the water, equilibrium can exist. However it is unstable. Suppose a small disturbance, such as shown in Fig. 7b starts. The gas pressure remains constant. However the water pressure at point (1) becomes less since there is a shorter column of water above it, and the pressure at (2) becomes greater than the gas pressure. Thus the disturbances at (1) and (2) grow. The equilibrium is unstable.

Another example is the "kink" instability in a linear pinch (a straight-line plasma column carrying a current). The current produces a magnetic field encircling the plasma column. If a small kink should start, it will grow, because the kink causes the field strength to increase on the inside and decrease on the outside. The plasma column moves toward the weaker B field. See Fig. 8.

Many specific types of macroscopic instabilities have been classified, and overcome or reduced. A general principle has been established that a plasma will be macroscopically stable if it is in a region of minimum magnetic field strength. That means that the magnitude of B must increase in every direction moving away from the plasma. Unfortunately no configuration that is both "minimum B field" and closed (without end leakage) has been developed. It appears that if particles are on the average in minimum magnetic field regions, stability can be achieved. The full story of all of the instabilities and all of the field configurations which have been tried is too long for here; see the References.

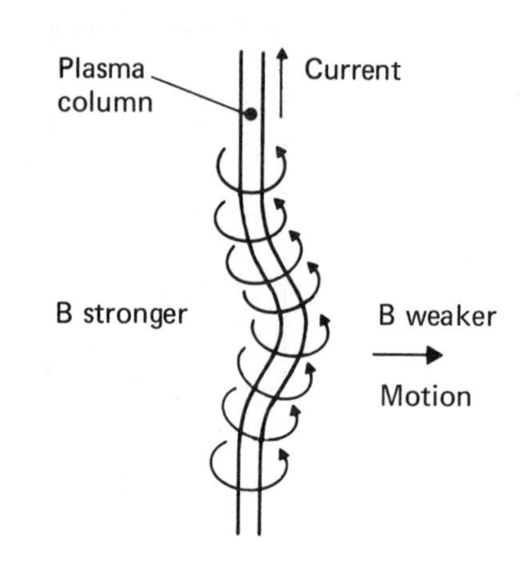


Fig. 8. The Kink instability.

The microscopic instabilities involve the individual particles considered on a statistical basis, and are much more difficult to understand. They arise primarily because the particles are not in thermodynamic equilibrium. For example, in a magnetic mirror machine, the end losses mean that isotropy does not hold. Situations can arise that are analogous to a laser. In the case of a laser, some atoms are "pumped" into a higher energy state, so that the occupancy of this state is much greater than in equilibrium. The passage of light of the proper wavelength can trigger these atoms to drop to a lower energy level, essentially all at once. The similar situation in the case of a plasma can lead to the sudden loss of confinement. More details are given in some of the References.

Even if all instabilities can be overcome, confinement is not perfect. There is a slow diffusion of the plasma across the magnetic field, caused by collisions. Consider two particles, each spiraling around the direction of the magnetic field. If they collide the directions of their motion will usually change. Thus they start off on new helices. The average behavior of many collisions leads to a gradual spreading of the plasma, rather like the random walk problem.

Although magnetic field strengths on the order of 3 to 10 $\text{Webers/m}^2$ are needed, there is no problem in obtaining this size field. It is almost certain that superconducting wire will be used for the windings, since little energy will be needed to sustain the field that way. There is a problem with the self-stress of the coils. This stress is a simple result of the magnetic field produced by the coil acting on the current in the coil. Very large forces result. The strength of most materials is low and the materials are brittle at the low superconducting temperatures. Another major problem concerns the magnetic energy stored in the fields and what would happen to it if the circuit was suddenly interrupted by a failure. The result would be like a small lightning bolt. See Prob. 5.

## HEATING

The plasma not only has to be confined for a sufficient time and density, but it must be heated to a high enough temperature that the particles can approach each other close enough for fusion. Several schemes have been proposed. One uses the heating effect of a current passing through a resistance (ohmic heating). We have seen that an induced current helps the confinement problem. It can also heat the plasma. Several difficulties arise. As the plasma temperature rises, the resistance goes down, so the heating, which goes as $i^2R$, becomes more difficult. Also there is an upper limit to the circulating current before instabilities set in. Finally the induced current exists only while the magnetic flux is changing. Since there is a limit to the maximum flux, the induced current can be of only limited time duration. Such a reactor could operate only in a pulsed mode. Another method of heating relies on the temperature rise associated with adiabatic compression. Instead of material pistons, the magnetic field strength is increased to compress the plasma. The pinch effect also produces compression. Again this is suitable only for pulsed operation. High speed neutral particles (which are unaffected by the magnetic field) can be injected. Once inside the plasma they collide, giving up some of their energy, become ionized, and trapped by the magnetic field. Other ideas consider the use of induced turbulence or waves which can transfer energy to the plasma. Heating is made difficult by the general fact that as temperature rises, the rate of loss of energy increases. One mechanism of energy loss in a magnetically confined plasma is radiation by the electrons. Since the electrons are spiraling, they are accelerating. Any time a charged particle is accelerated it will radiate electromagnetic waves. The radiation from the positively charged particles is negligible compared to that from the electrons, since the electrons have higher speeds and smaller radius orbits. See Prob. 6. Heavier impurity atoms also greatly increase the radiative losses, since they are not completely ionized. The remaining electronic levels can be collisionally excited, and the resulting radiation is in the x-ray region.

The best present magnetic confinement scheme (the Tokamak) achieves about a tenth of what is needed in all of the important parameters: ion density, confinement time, and ion temperature. It is hoped that a larger machine with higher magnetic field strengths will bring all of these up to the desired level.

## INERTIAL CONFINEMENT

The other main confinement approach relies on the fact that it takes time for a plasma to expand. Suppose a solid particle is $10^{-3}$ m in diameter, and is suddenly heated to $10^8$ °K. The average speed of hydrogen at that temperature is about $10^6$ m/sec. Thus it will take $10^{-3}/10^6 = 10^{-9}$ sec for a particle to travel the diameter. The Lawson criterion requires the product of density and confinement time to be $10^{14}$ sec/cm$^3$. Thus for $10^{-9}$ sec, the densities of the order of $10^{23}$/cm$^3$ are required. This is the order of magnitude of densities of solids and liquids.

The scheme then is to form a small drop of the fuel and suddenly heat it by lasers or energetic electron beams. A "microexplosion" or "mini-H-bomb" results, with an energy release perhaps on the order of one pound of TNT. The reaction would take place in a vacuum chamber perhaps 1 m in diameter. The pressures after expanding that much would be reasonable. See Prob. 4. The main problem at present is in obtaining laser or electron beams of sufficient energy.

REFERENCES


1. F. I. Boley, Plasmas-Laboratory and Cosmic (Van Nostrand, 1966).

2. R. F. Post, "Fusion Power", Scientific American, Dec. 1957.

3. T. K. Fowler and R. F. Post, "Progress toward Fusion Power" Scientific American, Dec. 1966.

4. W. C. Gough and B. J. Eastlund, "The Prospects of Fusion Power", Scientific American, Feb. 1971.

5. F. F. Chen, "The Leakage Problem in Fusion Reactors", Scientific American, July, 1967.

6. M. J. Lubin and A. P. Fraas, "Fusion by Laser", Scientific American, June 1971.

7. B. Coppi and J. Rem, "The Tokamak Approach to Fusion Research" Scientific American, July 1972.

8. L. Wood and J. Nuckolls, "Fusion Power", Environment, May 1972.

9. W. D. Metz, "Laser Fusion," Science 177, 1180 (29 Sept. 1972).

10. R. F. Post, "Prospects for Fusion Power," Physics Today 26, No. 4, 30 (Apr. 1973).

11. L. Spitzer, Jr., "The Stellarator," Scientific American, Oct. 1958.

12. D. J. Rose, "Controlled Nuclear Fusion: Status and Outlook," Science 172, 797 (21 May 1971).

PROBLEMS

1. How much energy is needed to bring two protons within $2x10^{-15}$ m of each other? [This is about the distance at which nuclear forces become effective.] Express your answer in eV. Because of a quantum mechanical effect called "tunneling," the reaction actually proceeds with reasonable probability at less than one tenth of the energy calculated here.

2. Use Eqns. 1 and 4 to derive an expression for the average size of the orbits as a function of temperature in a uniform magnetic field. How does $\overline{v}_\perp$ compare to $\overline{v}$ for random orientations? Evaluate your expression for electrons and deuterons (heavy hydrogen, or $H^2$) for temperatures in the fusion range and magnetic fields around 1 - 4 Webers/$m^2$.

3. Why can one not confine a plasma with a static electric field?

4. Assume a liquid drop $10^{-3}$m in diameter is heated to $10^8$ °K, and then allowed to expand in a vacuum to a diameter of 1 m. What is the final pressure? Assume ideal gas behavior. Hint: recall the behavior of an ideal gas undergoing free expansion. The density of liquid deuterium is .1739 g/$cm^3$.

5. Calculate how much energy is stored in a magnetic field with the following parameters: $B = 5$ Webers/$m^2$, and the shape is a cylinder 3 m in radius and 10 m long. If this energy is released in 0.5 sec, what is the average power?

6. Find the ratio of the centripetal acceleration of an electron to that of a deuteron. Hints: How does $\bar{v}$ depend on mass? How does radius depend on mass and speed?

# Unit 21
# Energy Distribution

## INTRODUCTION

It is clear that the U.S.A. has an energy problem. However, merely producing adequate energy is not sufficient. It must also be transported from the production point to the location to the consumer. We must consider not only the case of taking fossil or nuclear fuels from the mine or well to the consumer, but also the transmission of electricity and synthetic fuels. Most of the future electric power plants will probably be located a long distance from metropolitan areas. In the first place there is little land left on which to build new plants. Then there is the pollution problem. Air quality standards may not permit adding another power plant in a heavily populated area where the pollution levels are already high. Nuclear power plants should definitely be built far from any populous areas for safety reasons. Solar power plants need large land areas and are more effective in sunny regions (see Unit 22). The sunniest areas are in the southwestern states, far from the major population centers except for California. Fusion power plants must be located by the oceans in order to obtain their fuel (see Unit 28). Hydroelectric plants obviously must be located where there is a river with suitable surrounding topography.

The basic ways of transportating energy are as solid, liquid, or gaseous fuels, including man-made ones, or as electricity. It does not appear feasible to tranmit energy in large quantities in the form of electromagnetic waves, such as microwaves, at the present.

## SOLIDS

The basic solid fuels are coal, which is our most abundant fossil fuel, and the fissionable elements uranium and plutonium. A possibility which is being considered in connection with solar energy is aluminum. Solar-produced electricity can be used to make aluminum from aluminum oxide. The metal is then shipped and/or stored until it is needed and then burned. The aluminum oxide is shipped the other way for reprocessing, so there is a closed cycle and no material resources are used up.

The main ways that solids are transported are by boat and train. These two methods are reasonably inexpensive, both in terms of dollars and energy. However, the costs for transporting coal are comparable to those for transmitting electricity, so there is a trend towards putting the power plant at the coal mine and sending the electricity. There are limitations to this practice based on the shortage of water in the western states near many of the coal fields and problems with the deterioration of air quality in presently clean areas. Conveyor belts and making a coal-water slurry which can be pumped have been used for relatively short distances.

LIQUIDS AND GASES

These categories are lumped together since the methods are similar. The major transportation method on land is by pipeline, although trucks and trains are also used. Most of the energy needed in a pipeline is to overcome frictional losses due to the viscosity of the fluid. If the flow is laminar, which will usually be the case, then the governing relation is Poiseuille's law:

$$Q = \frac{\pi R^4 (p_1 - p_2)}{8 \eta L} \qquad (1)$$

Where $Q$ is the volume per second, $R$ is the radius of the pipe, $L$ its length, $\eta$ is the viscosity, and $p_1 - p_2$ is the pressure difference. Note the dependence on the fourth power of the radius, indicating a tremendous increase in flow as the size of the pipe is increased. Thus there is a real economy of scale in this case, since the cost of the pipe increases approximately as only the first power of the radius. Since the flow rate is inversely proportional to the viscosity, and since the viscosity of gas is much less than for a liquid, more gas flows for a given pressure drop. On the other hand the energy content of oil is much greater on a volume basis. See Prob. 2.

We can find how much power is needed to pump fluid at a given rate through a pipeline. We will suppose that the pipe is level (see Prob. 4 for a non-level pipe). The net work done in moving a volume $\Delta V$ when there is a pressure difference $p_1 - p_2$ is

$$\Delta W = (p_1 - p_2) \quad \Delta W$$

where we assume that the change in volume due to the change in pressure may be neglected. The volume $\Delta V$ is $Q \cdot \Delta t$, so

$$\Delta W = (p_1 - p_2)\ Q\ \Delta t$$

Since power is the time rate of doing work, we have

$$P = \Delta W/\Delta t = (p_1 - p_2)\ Q \tag{2}$$

If we solve Eqn. 1 for $p_1 - p_2$ and substitute into Eqn. 2 we obtain

$$P = \frac{8\eta L Q^2}{\pi R^4} \tag{3}$$

We see that it is not economical to try to pump too much through a given line. As the flow rate $Q$ doubles, indicating a doubling of the rate of energy transported, the needed power quadruples. It is better to increase the size of the pipe instead.

An especially attractive form of gaseous fuel is hydrogen. Hydrogen can be made from water by electrolysis or possibly by a direct solar process. Obviously some new primary energy source such as nuclear or solar energy must be used. Although losses are involved in the double conversion (electricity → hydrogen → electricity), there are several advantages. If inexpensive fuel cells can be developed much of the inefficiency of the conversion of hydrogen to electricity can be eliminated. Gas pipes are underground, while existing electrical transmission lines are mostly above. Because of the economy of scale implied by Eqn. 1, it costs about 20% as much to transport a given amount of energy in gaseous form as it does in the form of electricity if long distances are involved. Gas can be stored so as to even out the varying day-night demand for electricity. The storage feature is especially useful in the case of solar energy. Hydrogen can be burned rather cleanly in an automobile, whereas adequate storage batteries have not yet been developed. Hydrogen appears to be the best replacement for petroleum products for cars, trucks, and airplanes. It would probably be carried as a metal hydride in the first two cases, and as liquid hydrogen in the last case. Hydrogen is a recyclable fuel, with nature giving a free return. There are so many advantages to using hydrogen as the primary method of transporting energy that several writers (see Refs. 2 and 5) believe that we are headed towards a "hydrogen economy". For example, the only energy source one would have coming into a home would be hydrogen. A fuel cell would convert it to electricity, it would be burned directly for space and water heating, and one could even refill one's automobile by a simple connection at home. At present energy consumption rates, there would be an entirely negligible lowering of the ocean (about 0.02 mm) and increase in rainfall (about 0.7 mm per year). Many people suffer from the "Hindenburg syndrome" and fear explosions of hydrogen. The possibilities of explosion and fire are of course quite real; however, hydrogen is not significantly more dangerous than natural gas and can be safely handled.

The only practical way to move a fluid over water at present is by tanker. If a gaseous fuel is to be carried by ship it must first be liquified so that reasonable quantities can be carried without having high pressures. Some trial runs have already been made with liquified natural gas (LNG), and large scale use is planned. Accidental spills are the main problem with tankers. In the case of oil it is disasterous to birds and some marine life and renders beaches unusable. The main hazard with a LNG spill is fire.

## ELECTRICITY

The transmission of energy in the form of electricity has many advantages, and some disadvantages. There are no moving parts to wear out, little labor is involved, and electricity can be directly used for a wide variety of purposes. About 10% of the final-use energy is transmitted as electricity now. [Because of the inefficiency of producing electricity, the generation of electricity takes about 25% of the primary energy usage.] The main concerns with the transmission of electricity are the energy lost, and the esthetics of and land used by above-ground transmission lines.

The main loss in above-ground lines is the Ohmic loss, or Joule heating. The power loss is

$$P_{loss} = i^2 R_{lines}$$

while the power transmitted is

$$P_{trans} = i V$$

The resistance of the lines can be reduced by using low resistivity materials and making the lines large in diameter. There are clearly limits to how large the wires can be made. Power loss can also be reduced by increasing the voltage. The last two equations above can be combined to give:

$$P_{loss} = \frac{(P_{trans})^2 R_{lines}}{V^2} \qquad (4)$$

It is evident that for a given amount of power transmitted, the power loss varies inversely with the square of the voltage. High voltage transmission lines crossing the countryside are a common sight. The voltages used in the transmission lines range from around a hundred thousand up to 750000 V. These voltages are much too large for home use, so they must be reduced.

The ease of stepping voltage up or down with transformers is the main reason that alternating current is in common use. Transformers require a time-varying magnetic field, produced by the time-varying current in the primary coil, to induce a current in the secondary coil. Transformers can be built with extremely high efficiencies. The upper limit on voltage is set mainly by the ultimate "break-down" of insulators, including air, allowing the current to go where it is not wanted.

There are several advantages to using direct current for very high voltage transmission lines, which largely compensate for the difficulty in changing the voltage in the case of dc. Only two wires are needed, instead of the three needed for three-phase ac, thus saving wire cost. Direct current lines are a little easier to insulate, since the peak and the rms voltages are the same. In ac lines the peak voltage is about 40% higher than the rms voltage; one must insulate for the peak voltage. An alternating current is surrounded by a time-varying magnetic field, and this time-varying magnetic field will induce a current in the neighboring wires which is out of phase, thus reducing the transmitted power.

Another disadvantage of ac arises from what is called the skin effect. This is a concentration of the current in a relatively thin layer near the surface of a conductor which arises from the interaction of the time-varying electric and magnetic fields associated with an alternating current. For copper at 60 Hz, the skin depth is about 8 mm. The depth decreases as the resistivity goes down and the frequency increases, varying as the square root in both cases. Thus only the outer portion of copper wires larger than about 1.6 cm in diameter is effectively used, and therefore the resistance decreases only linearly with increasing diameter, instead of quadratically as happened for dc or smaller wires.

An additional problem with ac transmission lines concerns the interconnection of many generating plants for load-sharing and emergency back-up. If one of the plants shifts frequency or even phase a little bit it can cause surges in the system leading to a general power failure. Direct current transmission obviously does not have this problem.

There are two additional power losses which arise for ac lines in the case of underground transmission lines. The wires are surrounded by some insulating material and then by a grounded conductor. The capacitance of this coaxial arrangement is considerable. This capacitance is charged and discharged with every change of the cycle. Since the charging current is 90° out of phase with the voltage, there is no power loss associated with the charging of the capacitance. However, since the charging current has to flow through the conductor, additional Ohmic losses arise. The charging current loss can be reduced by adding inductors along the line. Another loss, called dielectric loss, is significant for many solid and liquid insulators (but small for gaseous insulators). The time-varying voltage produces a time-varying electric field, which causes the electric dipoles induced in the dielectric to be functions of time. Energy can be transferred from the electric field to heat. The dielectric loss increases with frequency and the square of the voltage. These two losses limit the voltage and length of underground ac transmission lines. If the line is too long, the maximum ratings of the line can be exceeded simply by the charging current and dielectric losses. For oil-paper insulation, this critical length for ac is about 30 to 50 miles. A gas-insulated cable can be perhaps 500 miles long. Direct current lines do not suffer from these effects, and a conversion to dc would more than double the power handling capabilities of existing underground transmission lines which are insulated with oil and paper.

Two desires, elimination of unesthetic above-ground transmission lines and the need for more efficient transmission of electricity, point toward the use of underground cooled lines. Cooling from ambient temperatures down to the boiling temperature of liquid nitrogen (77 °K) reduces the resistance of copper to about 15% of the normal value. Thus more power can be transmitted for the same loss. Dropping down to the boiling temperature of liquid hydrogen (20 °K) reduces the resistance by an additional factor of almost 250. Because the skin depth varies as the square root of the resistivity, a decrease in resistivity by a factor of 100 results in a decrease in effective resistance by only about a factor of 10 in the case of ac. So again there is an advantage to dc.

The real "magic" occurs when very low temperatures are used with certain materials, and the superconducting state is reached. In Type I superconductors the resistance is zero and there are _no_ losses. Unfortunately Type I superconductors can carry only moderate currents before they switch out of the superconducting state. Type II superconductors, on the other hand, can carry much larger currents. There are some losses in a Type II superconductor, based on the magnetic field being inside the material in quantized "fluxoids". With alternating currents the fluxoids are moving about and interact with imperfections in the superconductor, leading to losses. These losses increase linearly with frequency and as the third or fourth power of the current. Even with direct current, there is still some small loss associated with "slippage" of the fluxoids. There is also the cost disadvantage in an ac to dc conversion and back. Some recent work suggests that by alternating layers of superconducting material with nonsuperconducting material, the losses may be kept quite low. There is a real danger associated with superconducting transmission lines if they are allowed to warm up above the superconducting temperature, perhaps by the failure of a refrigerator. The proposed currents are large enough that the Joule heating associated with the normal conducting state would be tremendous. The heat could transfer rapidly enough to the surrounding cooling fluid that an explosive expansion would result.

REFERENCES

1. D. P. Snowden, "Superconductors for Power Transmission" Scientific American, April 1972, p. 84.

2. D. P. Gregory, "The Hydrogen Economy" Scientific American, Jan. 1973, p. 13.

3. J. A. Fay and J. J. MacKenzie, "Cold Cargo" Environment, Nov. 1972, p. 21.

4. W. D. Metz, "New Means of Transmitting Electricity: A Three-Way Race" Science, Dec. 1, 1972, p. 968.

5. T. H. Maugh II, "Hydrogen: Synthetic Fuel of the Future" Science, Nov. 24, 1972, p, 849.

6. P. H. Rose, "Underground Power Transmission" Science, Oct. 16, 1970, p. 267.

PROBLEMS

1. Another alternative for the transmission of power would be by mechanical means. Investigate the feasibility of using a rotating rod. How big a rod would be needed to transmit 1000 MW? See Unit 7 for information on maximum rotational rates, and look up the maximum torsional strength.

2. Compare the rate of transmission of energy by pipelines carrying natural gas and oil. Assume the same size pipe and the same power used for pumping. You will have to look up the energy contents (see Appendix 4) and the viscosities for both gas (methane) and oil.

3. Why will the Alaskan pipeline be heated?

4. How much more power will be needed to pump a fluid uphill than that given by Eqn. 3?

5. Evaluate Eqn. 3 for several cases.

6. Design a transmission line for 500 MW of electri-

city. Assume a copper line, dc (which means two wires), a length of 500 miles, and a power loss of 3%. If the radius is 2 cm, what is the needed voltage?

7. The greater the temperature difference, the less efficient a refrigerator works. Assume an ideal refrigerator, and find how much work must be done to remove one unit of heat from 77 °K, 20 °K, and 4 °K. Actual refrigerators are only about 1/3 as efficient as the ideal. Is the saving in energy loss produced by cooling to 77 °K and to 20 °K greater than the additional energy required to produce the cooling? Assume that about 3% of the transmitted energy is lost by Joule heating at normal temperatures, and that the refrigerators must remove only the heat produced by the Joule heating (that is, neglect any heat leaks).

# Unit 22
# Solar Power

## EXTENT OF AVAILABLE POWER

With the end of fossil fuels definitely in sight (see Appendix 3), we need to find a new major long-term source of energy. Solar energy clearly is a long-term resource: it will be available as long as the earth is habitable. The only questions are the amount available and the technology for capturing it.

The energy flux from the sun is $1400\ W/m^2$ at the top of the earth's atmosphere (see Prob. 1). About 2% is absorbed high in the atmosphere, and 30 - 40% is reflected or scattered back into space by clouds and the air. Another 20% is absorbed by the air, leaving about 40 - 50% to be absorbed by the earth's surface. Thus there is about $630\ W/m^2$ at the earth's surface. This is an average figure. The flux is higher where there are no clouds and lower with clouds. We will use the average figure. Note carefully that the area used in defining the flux is measured perpendicularly to the direction of radiation. If the sun is not overhead, the projected area on the ground is larger, and the power per unit ground area is correspondingly reduced. The intensity on the ground is reduced by cos $\theta$, where $\theta$ is the angle of the sun from overhead. For example, if the sun is 45° away from the zenith, the power per unit of surface area on the ground is reduced to 445 $W/m^2$. This effect is especially important during early morning, late evening, and winter months for locations well away from the equator. We will use 445 W per square meter of ground as an average value for the middle 8 hours of a day and locations not too close to the poles. See Prob. 2.

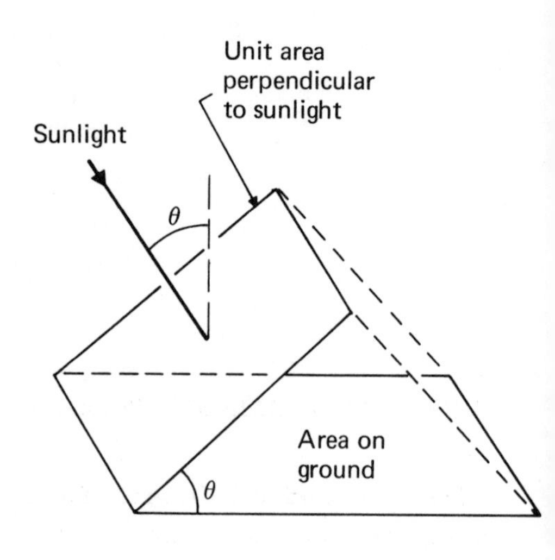


We will assume that 15% of the ground-level solar energy can be converted to electricity, allowing for various losses including the thermodynamic one, and that solar energy is available 20% of the time, allowing for night and bad weather. The average electric power available is thus 13 $W/m^2$. This figure is conservative.

We next need to know how much ground area is available for solar energy purposes. The total land area of the world is about $1.5\text{x}10^{14}$ $m^2$. About $5\text{x}10^{12}$ $m^2$ are needed for housing and industry, $5\text{x}10^{13}$ $m^2$ for food production, $7\text{x}10^{13}$ $m^2$ are unsuitable (too far north or south, or too mountainous), leaving potentially $2.5 \times 10^{13}$ $m^2$ which could be pressed into solar energy production.

The total potential electrical power is thus:

$$(13\ W/m^2)(2.5 \times 10^{13}\ m^2) = 3.25 \times 10^{14}\ W$$

The present world power consumption is about $6.5 \times 10^{12}$ W. There is thus ample power for present needs.

What about future needs? The most rational way to approach this is to note that world population must level off, and $16 \times 10^9$ people is a reasonable figure, limited primarily by food and water. We will also suppose that the per capita power consumption levels off at about twice the present U. S. rate, or at $2 \times 10^4$ W. This leveling off is necessitated by pollution and possible climatic effects. The total reasonable sustained world power consumption is thus:

$$(2 \times 10^4\ W/person)(16 \times 10^9\ people) = 3.2 \times 10^{14}\ W$$

[The near agreement of the above two figures is a coincidence, in the sense that the data were not juggled or manipulated with that end in mind.]

As an alternative confirmation that the potential supply of solar energy is quite adequate, let us see how much land area is needed to supply all of the present U. S. power needs, which is about $2.2 \times 10^{12}$ W. At $13\ W/m^2$, $1.7 \times 10^{11}\ m^2$ are needed. This is about 65000 $mi^2$ or an area about 255 mi by 255 mi. This area is only about 2% of the land area of the "old 48" states (as a comparison, roads use up about 1.5% of our land area), or 57% of the area of Arizona. We pick on Arizona since it is the sunniest state. As an aside, 44% of Arizona is federally owned now.

## METHODS FOR CONVERTING SOLAR ENERGY

Perhaps the first thing to come to mind is the use of "solar cells" or photovoltaic devices, which have been used so successfully in the space program. The principal problem is that they are more than 100 times too expensive at present for large scale ground use. It is unlikely that the price will be reduced sufficiently in the near future.

Another approach uses solar energy to decompose water into hydrogen. This can be done with the aid of a catalyst or by a multistep chemical reaction. High temperatures are needed. See Refs. 3 and 5 for more discussion of this method.

The method which is receiving most attention now is a thermal cycle in which steam is generated and conventional turbines and generators are used. For reasonable efficiencies and low thermal pollution high temperatures are needed.

## OBTAINING HIGH TEMPERATURES

The temperature of an object rises as long as the rate that energy is received exceeds the rate of loss, and stabilizes when the rates are equal. The rate of energy loss rises with temperature, making high temperature difficult to achieve. The difficulty with solar energy is that the rate at which energy arrives per unit area is very low, around 500 $W/m^2$. For comparison a home furnace may have about $10^5$ $W/m^2$ in the flame. One way around this is to concentrate the sunlight to a higher flux by lenses or mirrors. Another way is to work hard at reducing the rate of energy loss.

Optical concentrators can either bring the sun's energy into an approximate point focus or a line focus. In the former case one obtains the highest degree of concentration, but accurate steering mechanisms are needed to follow the sun's apparent motion. In the later case steering is not so critical and continuous steering possibly can be dispensed with. At this stage of development it is not clear whether mirrors or lenses are better. There are unanswered questions about how they hold up for long periods of time in the harsh desert environment. A major disadvantage of optical concentrators is that they do not work when the solar radiation has been diffused by cloud cover. Since cloudy conditions prevail in many parts of the world this is important. There are sufficient sunny regions, however, that this method should be developed vigorously.

The ways by which the solar collector loses heat are conduction, radiation, and of course the removal of energy to feed to the boiler. The latter can easily be set at any desired level. Conduction to the air can be eliminated by evacuating the region around the collector. The conduction losses along the pipes which circulate the operating fluid can be made small.

Radiative losses, however, can be quite important. The rate of radiation of energy for a quasi-blackbody is:

$$P = A\, e\, \sigma T^4 \qquad (1)$$

where $A$ is the area, $e$ is the emissivity, $\sigma$ is the Stefan-Boltzmann constant ( $= 5.67\times10^{-8}$ W/m$^2$°K$^4$), and $T$ is the temperature on the Kelvin scale. The emissivity is 1 for a blackbody and decreases as the body becomes a poorer absorber.

Radiation losses are automatically reduced when

concentrators are used, since the area of the absorber is small and hence the power radiated is small by Eqn. 1.

We can calculate the concentration factor needed for a given temperature as follows. We assume that we want to limit the radiative losses to 20% of the incoming solar energy, and that other losses are negligible. The incoming power is

$$S\,A_1\,e_1\,t\,e_2$$

where $S$ is the solar power flux (about 630 W/m$^2$ in this case, since steering will be used), $A_1$ is the concentrator (lens or mirror) area, $e_1$ the efficiency of the concentrator, $t$ is the transmission of the glass which surrounds the absorber (necessary for the vacuum), and $e_2$ is the emissivity of the absorber. The radiative losses are

$$\sigma e_2\,A_2\,T^4$$

where $A_2$ is the area of the absorbing surface. If the loss is to be 20% of the incoming power, we have:

$$\sigma e_2\,A_2\,T^4 = .2\,S\,A_1\,e_1\,t\,e_2$$

If we suppose $e_1 = .9$, $t = .95$, and that $e_2$ is the same for incoming and outgoing radiation, we obtain:

$$A_1 = 5.26\text{x}10^{-10}\,A_2\,T^4$$

For a temperature of 800 °K, we have

$$A_1 = 215\,A_2$$

The concentrator area is over 200 times the absorber area. Other area ratios will result if different temperatures or different fractional losses are assumed.

On first consideration it would appear that the use

of absorbers without any concentrators would lead to temperatures which are too low. The incident solar power is

$$S A_2 t e_2$$

where the symbols have the same meaning as before. The radiative losses depend on the difference between the temperature of the radiator and the surrounding environment as follows:

$$\sigma e_2 A_2 (T^4 - T_{env}{}^4)$$

where $T$ is the absorber temperature and $T_{env}$ is the environmental temperature. For higher values of $T$, one can neglect the term involving $T_{env}$, as was already done. To find an upper limit to the temperature obtainable, we suppose that no heat is removed for useful purposes:

$$S A_2 t e_2 = \sigma e_2 A_2 (T^4 - T_{env}{}^4)$$

If $T_{env}$ = 293 °K, we obtain $T$ = 349 °K = 76 °C, which is too low for power generation. The above calculation supposed that the emissivity was the same for absorption and radiation, which is true for a blackbody or a "graybody." Fortunately there is a clever "trick" which can be done.

The incident solar radiation and the outgoing thermal radiation from the absorber are largely in different spectral regions. See. Fig. 1. The wavelength of the maximum of the distribution for a quasi-blackbody is related to the temperature by Wien's law:

$$\lambda peak = \frac{2897}{T} \tag{2}$$

where $\lambda$ is in microns and $T$ in °K. The temperature of

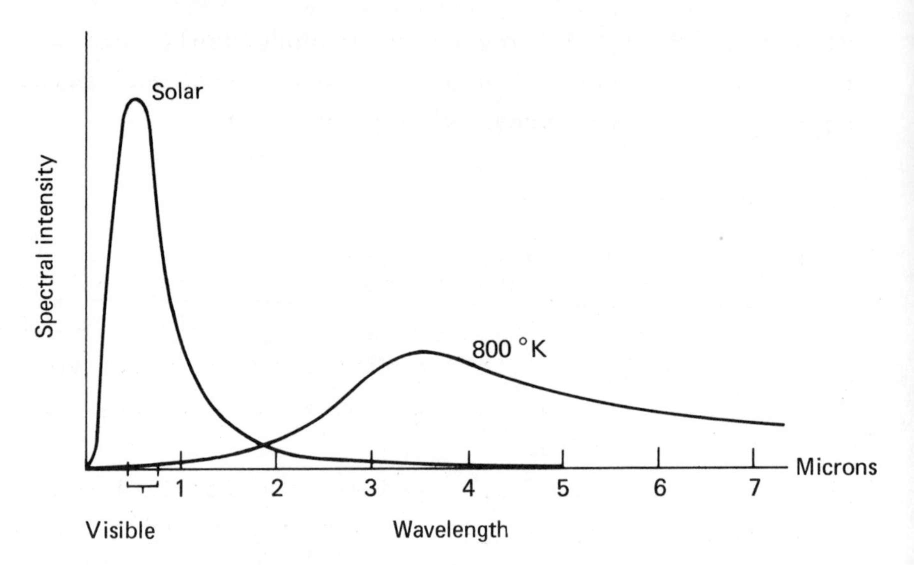


Fig. 1. Spectral distribution of solar and 800 °K light. The vertical scale is arbitrary, and not the same for the two curves. On the same scale the solar curve is always above the 800 °K curve.

the sun is about 5600 °K, giving a peak at .52μ (in the visible). If the absorber is at 800 °K (527 °C), the peak of the radiation from it is at 3.6μ (in the infrared).

The trick, then, is to have a surface which is a good abosrber in the .3 - 1μ region, and a poor absorber (and thus poor emitter) in the 2 - 10μ region. Fortunately some substances behave this way. This selective behavior can also be achieved by interference filters. See Refs. 1 and 2 for more details. At the present these special surfaces are expensive, but it appears that large scale techniques can be developed to make them at reasonable cost. The chief advantages of the selective surfaces are that they do not have to be steered and they function reasonably well on cloudy days. Even with heavy clouds an appreciable amount of solar energy reaches ground level. Perhaps a combination of concentrators in sunny areas and selective surfaces in cloudy areas will prove best.

## STORAGE

The most obvious thing about solar energy is that the sun shines only part of the time at a given point. Barring a world-wide highpower distribution network, which would have problems when the sun is over the Pacific Ocean, some sort of storage is needed.

Suitable sites for pumped-storage facilities (see Unit 9) are too limited for this method to be of general use. Sufficient storage by a flywheel seems unlikely (see Unit 7). Thermal storage appears feasible. The energy can not be stored as hot gases; too large a volume and too high pressures are involved. A well-insulated tank of some hot liquid, perhaps molten sodium or some molten salt, seems entirely possible without incurring too large an expense for insulation. There is a real economy of scale here, since the surface-to-volume ratio decreases as volume increases. The use of the latent heat of the liquid-to-solid transition has been suggested. Noneutectic mixtures of metals, which effectively have a broad temperature range associated with the transition, rather than a unique melting temperature, may be very suitable.

Another general approach is to store the energy chemically. Certainly if the processes to produce hydrogen from water directly with solar energy succeed, there would be an easily stored energy source. It might be feasible to use the electricity produced by a thermal cycle to make hydrogen from water and then store the hydrogen. Other chemicals, such as aluminum, which could be produced with the excess electricity during the day and then burned at night have been suggested. One advantage of a chemical storage method is that the chemical can be shipped to other locations, rather than having to transmit electricity over long distances. See Unit 21.

A difficult question in designing energy storage facilities is to decide on how much storage capacity is needed. Suppose three days of bad weather were taken as the design criterion. What happens if an unusual storm lasts four or five days? No matter what length of storage time is considered, the possibility of bad weather lasting longer than that exists.

## HOME USE

About 10% of our total energy consumption is used for home heating. In many climates at least a large portion of the heating could be provided by solar energy. The technology for heating is considerably simpler than for producing electricity. Temperatures under 100 °C suffice, so it may be possible to use flat absorbers without either optical concentrators or selective surfaces. Another advantage is that no further conversion to electricity or other forms of energy is needed.

Basically all that is needed is a large black surface with water tubes running along it. The surface should be enclosed with glass on the front and some insulation on the other sides. The glass will cut down on convective losses and provides a modest amount of selectivity since infrared radiation is not readily transmitted by glass. Evacuation of the region around the collector may be needed to reduce conductive losses.

How large a collector is needed? As an example, a 1500 $ft^2$ ranch house in Colorado in December needs about $1.25x10^9$ J/day of heat. We will assume that effective collection lasts 6 hours per day. The rate of energy absorption during that time must be

$$\frac{1.25x10^9 \text{ J}}{6(3600)\text{ sec}} = 5.8x10^4 \text{w}$$

If 50% of the 445 $W/m^2$ which reaches the ground can be utilized, we find that the area must be

$$\frac{5.8x10^4}{.5(445)} = 260 \text{ m}^2 = 2800 \text{ ft}^2$$

This is about double the floor area of the house. Since the collecting area should be inclined so as to be approximately perpendicular to the sun's rays, the collector could be built into the roof provided some overhang is used. For a house at 40° N lattitude, a roof at about 50° from the horizontal would provided good average winter interception. Clearly house design would have to be nontraditional.

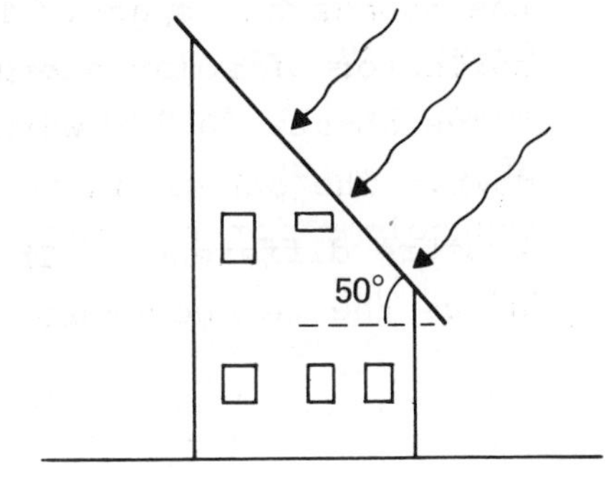


Storage is not a bad problem. Let us assume we need 48 hours of storage, which means $2.5x10^9$ J = $5.97x10^8$ cal. If a 10 °C drop in water temperature is tolerable, we need about $6x10^7$ gm of water. This is about a volume of 2100 $ft^3$, or about 16 x 16 x 8 ft, which could easily be contained in the basement.

Since there are no off-the-shelf collectors and most homes are not designed with solar energy in mind, the cost of solar heating is rather high--perhaps $2000 to

$5000. Since a backup heating system would be needed for unusually severe weather, this is an additional cost.

The technology is more difficult if solar energy is to be used for air conditioning. Rather paradoxically higher temperatures (around 125 °C) are needed for cooling by a thermal cycle. The principle is similar to the gas refrigerator. Probably some sort of selective surface will be needed. It is unlikely that home production of electricity from solar energy will be practical in the near future.

## OTHER POSSIBILITIES

So far was have considered the collection of solar energy only on land. Two other possibilities have been suggested: the oceans and space. The surface of the ocean is a good collector of solar energy, resulting in surface temperatures about 20 °C warmer than those at about 1000m down. The scheme is to run a heat engine on this temperature difference. The ideal efficiency is very low, since the termperature differential is small:

$$e = 1 - T_c/T_h = 1 - 278/298 = .067$$

Even so, since the oceans are so large, a considerable amount of energy is available. Clearly some working fluid other than steam must be used. One of the major difficulties is in transporting the energy back to shore. This might be accomplished by converting water to hydrogen.

Space has the advantage that it is never cloudy there. If the collector was in a synchronous orbit, it would be eclipsed by the earth for only a brief period each day. Power would be beamed down by microwaves which can penetrate clouds. The expense of such a venture is completely out of the question for now.

## POSSIBLE ADVERSE EFFECTS OF SOLAR POWER

A potential adverse effect of solar power is the possibility of altering the climate. Large scale air motions transport solar energy from the equator to the poles. This energy transfer keeps much of the earth at a moderate temperature. Without it the tropics would be too hot and the more northern and southern parts too cold. If large scale solar energy converters captured a significant portion of the solar energy in the tropics, and this energy was transported by wires or as hydrogen to the more northerly and southerly portions of the globe, these large scale air currents might be affected. This effect would be important only if energy consumption is 100 to 1000 times higher than it is now. The consequences are not clear at present. If the collectors had a higher absorption than the ground they replace so that less of the sun's energy was reflected into space there might be an increase in the average temperature of the globe. This effect could be alleviated by increasing the reflectivity of the ground at other places. Certainly many space stations beaming energy down to earth would have a warming effect. If we keep our energy demands to a reasonable level, it should be possible to avoid any serious climatic effects arising from the use of solar energy.

Taking solar energy from the oceans and redistributing the waste heat could upset the ocean ecology and possibly the ocean currents, which are also important for the climate.

Any time that electricity is generated by a thermal cycle there is the problem of thermal pollution. In most sunny regions water is scarce. However dry cooling towers can satisfactorily handle the waste heat problem.

See Unit 14.

REFERENCES

1. A. B. Meinel and M. P. Meinel, "Is It Time for a New Look at Solar Energy?" Bull. Atomic Scientists, Oct. 1971.

2. A. B. Meinel and M. P. Meinel, "Physics Looks at Solar Energy" Physics Today, Feb. 1972.

3. N. C. Ford and J. W. Kane, "Solar Power" Bull. Atomic Scientists, Oct. 1971.

4. A. L. Hammond, "Solar Energy: the Largest Resource" Science, Sept. 22, 1972.

5. B. M. Abraham and F. Schreiner, "A Low-Temperature Thermal Process for the Decomposition of Water" Science, June 1, 1973.

6. W. D. Metz, "Ocean Temperature Gradients: Solar Power from the Sea" Science, June 22, 1973.

7. C. Zener, "Solar Sea Power" Physics Today, Jan. 1973.

PROBLEMS

1. Show that the sun's power flux is $1400\ W/m^2$ at the earth from the following data: the sun's total output is $3.9 x 10^{26}$ W, and the distance is $1.49 x 10^{11}$ m.

2. Calculate in more detail the average reduction in solar power per unit of ground area as compared to the incident solar flux measured perpendicularly to the beam. Both daily and seasonal effects and the lattitude should be considered.

3. Why can't one overcome the effect of the reduction of power per unit area due to the inclination of the sun by having collectors equal in area to the ground area and tilting them so they are perpendicular to the sun's rays?

4. How much do radiative losses increase as the temperature goes from 400 °C to 500 °C?

5. How large a storage facility is needed for a 800 MW (electric output) power plant? Assume two days of storage, liquid sodium as the medium, a 35% thermal conversion, and that a 50 °C drop is permissible. If the energy losses are to be kept to 5% of the stored energy in two days and a single cylindrical tank is used, how thick must the insulation be? (See Unit 15.) Assume the temperature is 500 °C.

6. How fast must water by pumped through a home solar heating collector if a 10 °C temperature rise is desired? Assume the figures given in the text.

7. How long would a synchronous orbit space station be eclipsed each day? Hints: How wide is the earth's shadow? How far up is a synchronous satellite? How long does it take to go around once?

8. Can a solar-powered automobile be made? Use the power requirements given in Unit 5.

# Unit 23
# Air Pollution V:
# Possible Effects on the Climate

## INTRODUCTION

The two major concerns with air pollution are possible adverse effects on health, and changes in the global climate. We consider the latter here. The two extreme possibilities are that pollutants in the air may cause a general cooling, leading to another ice age, <u>or</u> a general warming which would melt the polar ice caps, causing a rise in the level of the ocean of perhaps a few hundred feet thus flooding many of the major cities of the world. An important point to remember is that the change in average global temperature from a glacial period to the maximum in between is only about 5 °C. The overall stability of the atmosphere is not well understood. There are several mechanisms which have a positive feedback effect: a small change produces effects which enhance the change further. Fortunately there are also some negative feedback mechanisms which lend stability.

## RADIATIVE EQUILIBRIUM

The overall average temperature of the earth is determined by the balance between the energy input, primarily from solar radiation, and the radiation of energy from the earth to space. To a reasonable approximation both the sun and the earth behave as quasi-blackbodies. The important relations for blackbodies are given in Unit 22, Eqns. 1 and 2.

The rate of radiation received by the earth from the sun can be written as

$$S \pi R_e^2 \alpha \tag{1}$$

where $S$ is the solar intensity at the earth, about $1400 \text{ W/m}^2$, $R_e$ is the radius of the earth, and $\alpha$ is the average absorption. This last quantity is sometimes written in terms of $a$, the albedo, or reflection, of the earth. The two are related by

$$\alpha = 1 - a$$

The geometrical factor in Eqn. 1 is the cross sectional area that the earth presents to the solar radiation. The same result may be obtained by integrating over a hemisphere, keeping in mind the inclination factor. See Prob. 4. The absorption depends on many things such as whether there is land, water, or snow, the type of vegetation and roughness of the land, angle of incidence, cloud cover, and so on. Equation 1 uses a global average value.

The rate at which radiant energy leaves the earth can be rather well approximated by blackbody radiation. In the spectral region involved the earth is indeed quite black, with $e \simeq .95$. The rate is

$$4 \pi R_e^2 e \sigma T^4 \tag{2}$$

where the terms have the same meaning as in Unit 22. Note that the geometrical factor is now the entire surface area of the earth, since all of the earth is radiating to space. In Expression 2, $T$ is an average effective temperature.

Radiative equilibrium means that the two rates, Expressions 1 and 2, are the same:

$$S \pi R_e^2 \alpha = 4 \pi R_e^2 e \sigma T^4 \tag{3}$$

or

$$T = (S\alpha / 4 e \sigma)^{1/4} = 280.3 (\alpha/e)^{1/4} \tag{3a}$$

where $S = 1400 \text{ W/m}^2$ was used. If the earth was a perfect blackbody, we would have $\alpha = e = 1$, and the temperature would be 280 °K. In actuality, $\alpha = .65$ and $e = .95$ are better values. The effective temperature is then about 255 °K. Clearly this is not the average ground temperature, but rather corresponds to the temperature some distance up in the atmosphere, for reasons given below.

The two factors in Eqn. 3a that can be altered by man are $\alpha$ and $e$. Some of the factors affecting the absorption are the cutting down of forests for farming land, the replacement of cultivated land by buildings and roads, changes in the amount of aerosols in the air, and changes in cloud cover induced by man's activities. Cloud cover can be altered directly by the injection of more water vapor into the atmosphere and by increases in particulate matter providing more condensation nuclei. Indirectly cloud cover can be altered by changes in the mean temperature which may be caused by other factors. The effect of aerosols on the absorption $\alpha$ is still disputed. One school of thought holds that particles will reflect sunlight back to space, thus reducing $\alpha$ and causing a cooling. The other points out that particles may directly absorb sunlight leading to a warming. It is gradually becoming clear that no simple statement can be made. The answer depends on the size and composition of the particles, and importantly on the absorption of the ground underneath.

The emissivity $e$ is rather close to 1 for many surfaces and is only slightly affected by man. Particulate matter falling on snow and making it dirty will increase $e$, for example.

The above discussion assumed that the only significant energy input was from the sun, which is correct at present. However, unconstrained growth in energy consumption by man (see Appendices 1 and 3) could lead to a significant energy input which would lead to an overall temperature rise. See Prob. 6.

Another important point to note is that Eqn. 3a refers to the effective temperature for radiating to space, not the ground temperature. Man's activities can also alter the relation between the ground temperature and the effective temperature. Thus the climate can change even when the effective temperature is not changed.

## ROLE OF ATMOSPHERIC GASES

The discussion of climatic effects is complicated by the differences in behavior of the atmospheric gases for incoming and outgoing radiation. Solar radiation lies mainly between .3 and 2 μ, with a peak at about .5 μ. Outgoing radiation lies in the infrared with a peak at (Eqn. 2, Unit 22)

$$\lambda_{peak} = 2897/255 = 11.4\ \mu$$

All of the major gases in the atmosphere are reasonably transparent in the region of the spectrum where most of the solar energy is located. On the other hand there are important absorption bands in the infrared, especially due to $CO_2$ and $H_2O$. The absorption curves for $CO_2$ and $H_2O$ are shown in Fig. 1. Thus (in the absence of clouds or particles) most of the incoming solar radiation passes freely through the air and reaches ground level. Most of the outgoing infrared radiation from the ground however is absorbed by the air even with clear skies. Only about 7% to 8% of the net outgoing thermal radiation from the earth reaches space directly from the ground. The rest is radiated from the air. A reasonable portion of the sunlight can diffuse through clouds and reach the ground. On the

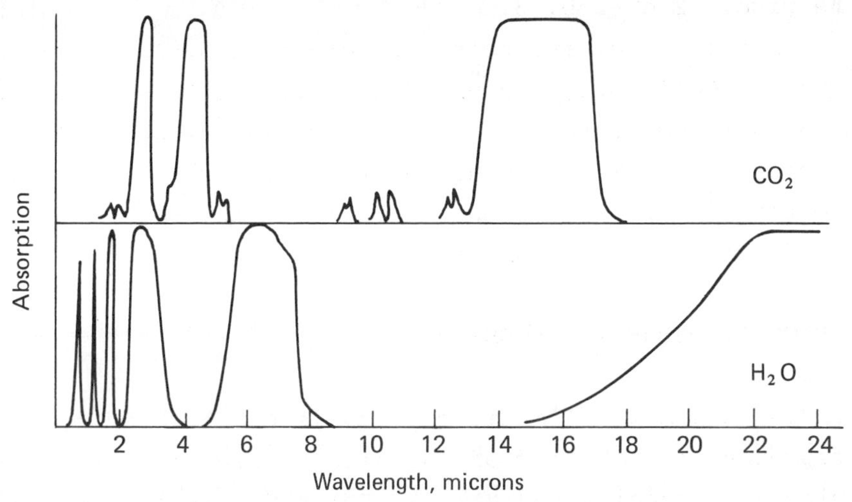


Fig. 1. Absorption of radiation by $CO_2$ and $H_2O$.

other hand clouds are essentially complete absorbers of infrared radiation, so the radiation to space again comes from the atmosphere rather than the ground. The effective temperature in Eqn. 3a thus largely corresponds to the average temperature of the region of the atmosphere radiating to space.

There is a considerable exchange of energy between the ground and the air, where conduction and latent heat are important as well as radiative transfer. The overall energy flow and transfer for incoming and outgoing radiation is shown in Fig. 2.

The various factors given above show that the ground is warmer than the radiative equilibrium temperature. One way to understand the increase in ground temperature caused by absorption of infrared by the air is as follows. The

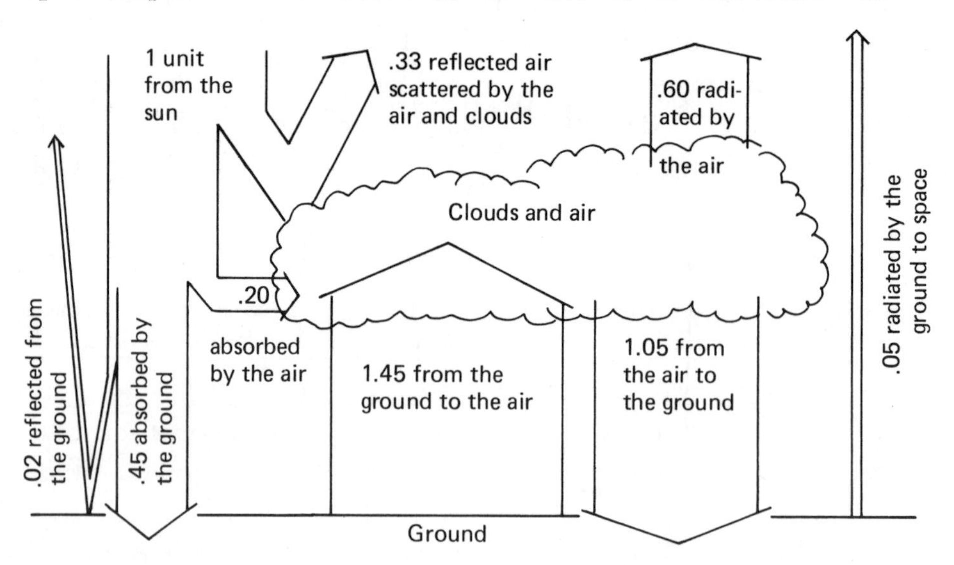


Fig. 2. The flow of energy to and from the earth. The figures are in terms of 1 unit of energy incident from the sun. The numbers are rounded off somewhat for simplicity.

presence of cloud cover and the absorption by the $H_2O$ and $CO_2$ prevent a considerable portion of the radiation from going to space. About half of the absorbed radiation is then reradiated back to the surface of the earth, warming it. A simple observation can underscore this concept. Cloudy nights are in general warmer than clear nights.

Another way to consider the warming effect is to start by noting that Eqn. 3 sets the effective radiative temperature. Whatever region is radiating to space must have this temperature. If the air were clear and transparent to infrared radiation, it would be the ground level which has this temperature. With absorbers in the air, most of the radiation to space originates from the atmosphere. This region then has the temperature required by Eqn. 3a. We saw in Unit 13 that normally the air temperature increases as one goes to lower altitudes. Thus the ground is warmer than the radiative temperature.

The warming of the ground caused by the absorption of infrared is often called the "greenhouse effect", since to some extent greenhouses are warmed by incoming sunlight passing through the windows, while the outgoing infrared radiation is partially blocked. It has been shown, however, that the main warming in a greenhouse comes from the prevention of convection of the air and of losses due to the evaporation of water.

Both $CO_2$ and $H_2O$ contribute to the absorption of the infrared, and hence to the warming of the ground. While more is absorbed by $H_2O$, it is the $CO_2$ which is of most concern. This is because the amount of $CO_2$ in the air is increasing as a result of burning fossil fuels, while the amount of $H_2O$ in the air is largely controlled by nature and only indirectly influenced by man. When the moisture in the air increases beyond a certain point, the excess rains out. The only changes man can make are by producing more condensation nuclei to hasten the rain, or by doing things which alter the average temperature. Temperature plays an important role since the saturation level of water vapor in air increases with temperature.

There is no automatic removal process for $CO_2$ corresponding to rain. About half of the $CO_2$ produced by burning fossil fuels remains in the air, the other half being absorbed in the oceans. The average concentration of $CO_2$ in the atmosphere has risen from about 280 ppm in 1860 to about 320 ppm in 1960, and is rising about 0.7 ppm per year now. If all of our fossil fuels were burned (which could happen within 100 - 200 years), and half of the $CO_2$ remains in the air, the concentration will be about 2000 ppm. See Prob. 5. The possibility of this much $CO_2$ seriously warming the climate is quite real.

Increasing the amount of $CO_2$ or $H_2O$ in the air will increase the ground level temperature first by blocking some more of the thermal radiation from the ground. This effect is limited, since now only about 7% to 8% of the net outgoing radiation reaches space from the ground (Fig. 2) and because there are still spaces between the combined absorption bands of $CO_2$ and $H_2O$ (Fig. 1). Thus certain wavelengths of the outgoing radiation are not absorbed. The main way in which an increase in $CO_2$ can increase the ground level temperature is by raising the altitude of the effective radiating region. A certain minimum amout of a gas, expressed in mass per unit area, is needed for good radiation. The relative density of gasses decreases with altitude. The amount of gas at a given altitude increases with the total amount of gas. Thus as the total quantity of $CO_2$ increases, the altitude above which there is sufficient

$CO_2$ for good radiation increases. Since the radiative temperature remains fixed and the lapse rate is a fixed number, an increase in altitude means that the ground temperature increases. See Prob. 3. The picture is more complicated for water vapor because of the change in temperature with altitude and the resulting condensation or solidification of the water.

Particulate matter in the air may also absorb the outgoing infrared radiation, further preventing direct radiation to space from the ground. Thus the role of aerosols is quite complicated, affecting both the incoming solar radiation and the outgoing thermal radiation. A satisfactory modeling of their role has not been done yet.

## SIZE OF THE EFFECT

There is no general agreement as to the increase in temperature if the $CO_2$ level doubles. The calculation is quite difficult, since it should consider a model of the global climate, with all of the feedback mechanisms, coupled to a model of the ocean. Present results range from about +2 °C to +5 °C for an increase to 600 ppm of $CO_2$.

Some of the difficulties in the calculation should be mentioned, especially those with a self-enhancing or positive feedback effect. We start by supposing that there is an increase in the $CO_2$ level and that a warming trend starts. We first consider the effect on the oceans. There is about fifty times more $CO_2$ in the oceans than in the air. As the temperature rises, the solubility of $CO_2$ in water decreases, resulting in the release of additional $CO_2$, which leads to additional warming, and so on. This is clearly a positive feedback loop. Another enhancing effect involves the water vapor. It is reasonable to suppose that the average <u>relative</u> humidity will remain the same. Thus the total amount of water vapor might increase as the temperature rises. Since $H_2O$ also has strong absorption bands in the near infrared region, there is an increased warming effect. As the air warms there is less snow and ice on the ground. Since they are good reflectors of radiation, a reduction in snow and ice increases the absorption and thus has an additional warming effect.

A negative feedback mechanism involving $CO_2$ concerns the polar ice caps. There is relatively little $CO_2$ in the ice. As a warming trend starts to melt the ice, fresh water with little $CO_2$ in it is released. This water can then start absorbing $CO_2$ from the atmosphere, reducing the warming effect and perhaps even reversing it. The polar ice caps then grow, $CO_2$ comes out of solution, and a warming starts. The period of this oscillation is about 50000 years, which is about the same as the period between the ice ages. It is tempting to speculate that the ice age cycle is in fact controlled by the $CO_2$ level, but this is not at all proven.

The role of clouds in the overall temperature change associated with an increase in $CO_2$ is not established. One line of reasoning says that an increase in temperature increases the air's ability to hold water and that therefore fewer clouds would form. The contrary argument states that an increase in temperature will cause more water to evaporate and increase the cloud cover. Furthermore, even the effect of a change in cloud cover is not certain. More clouds mean that more of the sun's energy is reflected back to space and less reaches ground level, which is a cooling effect. On the other hand, clouds are excellent absorbers in the infrared, thus blocking some more of the outgoing thermal radiation, which is a warming effect. More detailed calculations are needed to establish which effect dominates.

It is worth pointing out that the amount of plant life

has little long term affect on either the $CO_2$ or the $O_2$ in the air. To be sure as plants grow they take $CO_2$ out of air and put $O_2$ into it. But when they (or their leaves) decay, $O_2$ is taken from the air and $CO_2$ put into it.

Since we do not have a satisfactory model of the earth's climate and all of the factors affecting it, and since it is virtually impossible for man to remove a significant amount of $CO_2$ from the atmosphere, it would be prudent to reduce the formation of $CO_2$ as much as possible. Because of the positive feedback mechanisms, only a relatively small additional amount could trigger rather large changes. The $CO_2$ problem is another reason (besides the limited supply of fossil fuels) why alternative energy sources must be developed soon. It may be no exaggeration to say that the really significant side of the energy crisis is the production of $CO_2$ and the resulting climate modification.

REFERENCES

1. G. N. Plass, "Carbon Dioxide and Climate" Scientific American, July, 1959.

2. S. J. Williamson, Fundamentals of Air Pollution (Addison-Wesley, 1973). Ch. 4.

3. L. Hodges, Environmental Pollution (Holt, Rinehart, and Winston, 1973). Ch. 4.

PROBLEMS

1. How much does a temperature increase of 4 °C increase the radiation from the earth?

2. Find the change in the earth's effective radiative temperature if the following changes take place individually:
   a) the solar radiation increases 10%;
   b) the average absorption changes from .65 to .55;
   c) the emissivity drops from .95 to .90.

3. Suppose that an increase in $CO_2$ raises the effective height of the radiating layer in the atmosphere by 500 m. Since the temperature of the radiating layer stays about the same, what is the change in the ground temperature? Assume a lapse rate of 6 °C/km (see Unit 13).

4. Show by integrating over a hemisphere that the effective area for receiving solar radiation is $\pi R_e^2$ (instead of the geometric area of a hemisphere, $2 \pi R_e^2$). You will have to include the fact that the sun's intensity is measured in terms of an area perpendicular to the direction of the sunlight, not in terms of the area on the ground.

5. Verify the increase in the $CO_2$ concentration in the air if all of the fossil fuels are burned and half of the $CO_2$ remains in the air. You can use the coal reserves (Appendix 3), since it is the major source of fossil fuels. Assume the coal is pure carbon, and find how much $CO_2$ will be produced. There is about $2.4 \times 10^{15}$ kg of $CO_2$ in the air now.

6. Suppose that man's power production reaches 25% of the power absorbed from sunlight (which would happen

in about 330 years if exponential growth continues). Since this is no longer a negligible input, it must be added to the left-hand side of Eqn. 3. Show that the resulting radiative temperature is 270 °K assuming $\alpha = .65$ and $e = .95$.

# Unit 24
# Air Pollution VI: Detection of Air Pollution by Physical Methods

## INTRODUCTION

The goal of air pollution control is clean air, and various techniques for removing pollutants are obviously important. However, the measurement of pollutants is also a necessary part of control. It is necessary to know _what_ needs to be controlled before a control technique can be specified. Since some gases are harmless while others affect health adversely at very low concentrations, quantitative identification is needed. Air quality must be monitored continuously so that alerts may be given when severe levels are reached. In some areas the use of automobiles may be curtailed and industry shut down when the pollution is too large. Monitoring is needed to establish regional and global patterns for pollution. One region's or country's pollution may be visited upon another. There are questions about the rates of production of secondary (photochemical) pollution, and where the pollutants go when they disappear from the air.

Good detection techniques are needed for the enforcement of clean-air legislation. Unless there is some way to obtain firm evidence against a violator, a law is unenforceable. There is a bad handicap here in that there are definite restrictions about entering private property to obtain evidence against the property owner. Either some sort of "implied consent" law must be passed, or some rather sophisticated detection equipment must be developed that can unambiguously pinpoint the source. The latter is not simple. Support that a high level is detected just down wind from a given factory. The pollutants might have originated from the next factory over, and have looped to ground there.

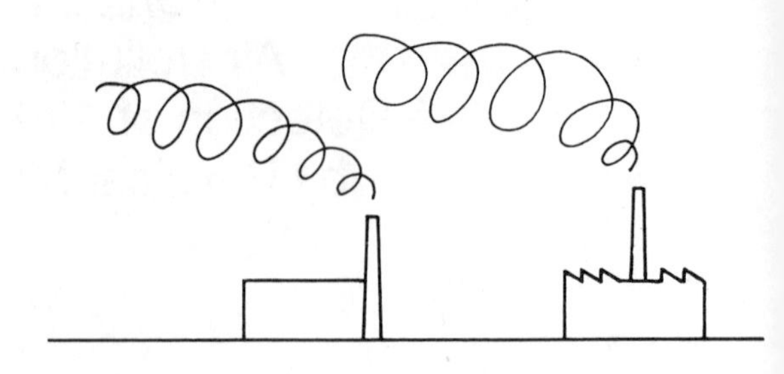

The traditional methods for detecting gaseous pollutants are based on wet chemical techniques. Air is bubbled through an appropriate solution, and then analyzed. The drawbacks include that it is not a continuous monitoring, and that many people have to be employed to collect the flasks and analyze the results. There is also the problem that, because of reactions in the solution, what is analyzed may not be what was in the air. For these reasons the trend is towards methods which allow a continuous unattended monitoring of air quality. Remote detection methods are obviously needed for global monitoring from satellites. We discuss here some of these methods which are based on physical principles.

## ABSORPTION SPECTROSCOPIC METHODS

Most of the physical methods are based on the fact that all atoms and molecules have quantized energy levels, with the set of levels unique to each. Thus when light is absorbed or emitted by a molecule, the energies of the photons are unique. Since energy of a photon is related to wavelength by

$$E = \frac{h\,c}{\lambda}$$

we have a unique set of wavelengths associated with the light absorbed or emitted by each kind of molecule. The specific wavelengths are often called "lines" because of the geometry of the usual spectroscope. These sets of lines are called the spectrum of the molecule. The emission spectrum in general has more wavelengths than the absorption, since the latter usually only involves transitions from the ground state.

Most molecules of interest in air pollution do not have any absorption lines in the visible portion of the spectrum. This is why they are invisible. The most useful set of lines lies in the infrared region. These transitions are associated with rotational and vibrational levels. There are a great number of transitions which have almost the same wavelength. Often at normal pressure the interaction of the molecules with each other causes the energy levels to shift sufficiently that the wavelengths merge into a continuous band. Even at low pressure the lines may appear to be merged together if the instrument has low resolving power. Thus one usually refers to a band spectrum in the case of molecules. Because the bands are reasonably wide, there is considerable overlapping of the bands of one molecule with those of other molecules. This leads to problems in identifying a molecule positively. However most pollutants of interest have one or more distinct bands.

Absorption spectroscopy is usually divided into dispersive and nondispersive methods. The former uses some method to separate light by wavelength, such as a grating or prism. A nondispersive instrument is set up along the following lines, where we use a CO detector as an example. Since no dispersing device is used, there must be some way to insure that only CO is detected. Figure 1 shows a schematic diagram. When a gas absorbs light it will heat up, resulting in a pressure rise if the volume is fixed. If there is the same amount of light energy in the wavelenghts corresponding to the CO bands in the upper and lower beams in the CO cells in Fig. 1, the pressure rise will be the same, and the pressure difference will be zero. However if there is any CO in the air in the sample cell, some of the light in the CO bands will be absorbed there, reducing the amount available to increase the pressure in the lower CO cell. The pressure difference thus depends (linearly for small atmospheric concentrations) on the CO in the atmosphere. The presence of the reference cell is to compensate for the infrared absorption by other constituents of air, especially $CO_2$. It is necessary to remove the water vapor from the sample. The presence of water vapor, which absorbs

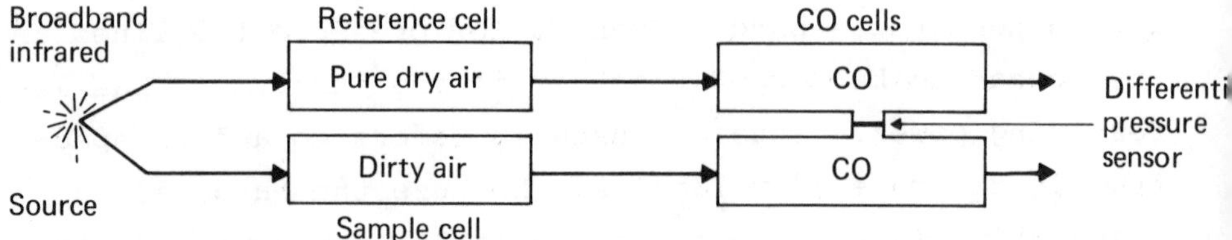


Fig. 1. A nondispersive CO detector.

readily in the infrared, could not be compensated by a fixed reference cell, since humidity varies greatly. The main problem with this method is that other pollutants which may be present simultaneously with the CO can also absorb in the infrared. Since their presence is not compensated by the reference cell, erroneous readings could result if any of their bands overlap those of CO.

Dispersive instruments, which separate the light according to wavelength, can in principle eliminate the interference problem, since the complete set of bands is unique. However in practice it is not quite so easy since only a limited range of wavelengths can be covered by a single instrument and the bands are often almost the same. One of the forms of dispersive instruments which utilizes information from many bands simultaneously is called a correlation spectrometer, shown schematically in Fig. 2. There is a broadband light source, the sample, a grating or prism to separate the wavelengths, the mask, and a detector. The mask is the important feature. It allows light of only those wavelengths which correspond to the absorption bands of the particular pollutant to pass. If the mask is displaced from its normal position, light from adjacent spectral regions where there is little or no absorption can fall on the detector. When the mask is in the normal position and there is strong absorption

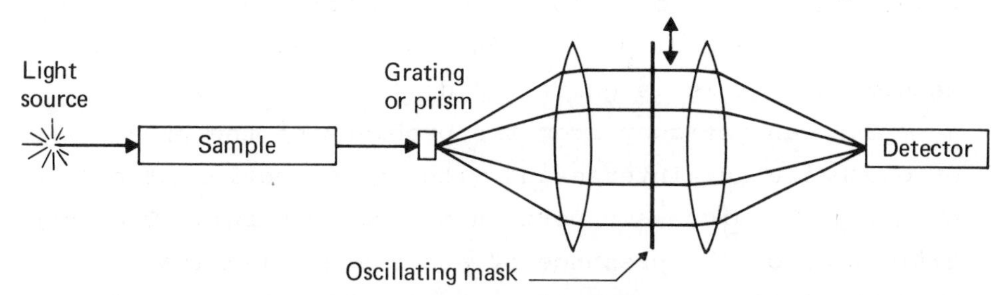


Fig. 2. Schematic of a correlation spectrometer.

by the pollutant, little light reaches the detector. Thus when the mask is oscillated back and forth, there will be a varying signal from the detector whose strength increases when more of the pollutant is present. Because many bands are used simultaneously there is little interference from other pollutants. In some cases it may be desirable to eliminate certain bands from the mask when they overlap too much with those of other gases likely to be present.

One interesting point about absorption spectroscopy is that the light source and sample region need not be within the instrument. A distant light source, even the sun, and a long path in the air can be used. Thus this method is suitable for use in aircraft or satellites. An average concentration along the path is obtained. Because a long sample region is possible, low concentrations can be measured.

Another variation on spectroscopic methods uses what is called Fourier transform spectroscopy. The basis of this method is something like a Michelson interferometer in which one mirror is moved back and forth so that the path difference changes. As a result the interference pattern changes. The variation in the interference pattern is basically a Fourier transform of the spectrum. The spectrum can be reconstructed by computer. However, since the transforms are also unique to each substance, identification can be made directly. The chief advantage of this method is that it is quite rapid.

One can get away from the problems of the presence of one pollutant interfering with the detection of another by using the microwave portion of the spectrum. The identification of the presence of atoms and molecules by certain characteristic microwave frequencies has been used for a number of years in radio astronomy to detect molecules in outer space. The principle difficulties with using microwaves for air pollution detection are the low sensitivity, and the reaction of the pollutants on the metal walls of the microwave cavity. Progress is being made to overcome the difficulties.

CHEMILUMINESCENCE

In some cases the reaction of one molecule with another leaves one of the products in an excited state. The molecule de-excites itself by emitting a photon, again of a definite wavelength. Because of the combination of a specific chemical reaction followed by a photon of a definite wavelength, this method shows very little interference from other co-existing pollutants. The main form of interference comes from other molecules which can collisionally de-excite the molecule in question, leading to an indication which is too low. One example is

$$NO + O_3 \rightarrow O_2 + NO_2^*$$

where $NO_2^*$ means that the $NO_2$ is left in an excited state. This reaction can monitor NO or ozone ($O_3$). This process can also be used for $NO_2$ by first chemically converting it to NO. Another reaction involves ozone and etheylene. Aside from the need to supply the reactant on a continuous basis, this method has the disadvantage that few pollutants have suitable reactions. It appears that chemiluminescence can be applied to the detection of $SO_2$ and peroxyacetly nitrate (PAN). An atomic oxygen source, which is not easy to make, is needed for the former, however.

The flame photometric detector is related to chemiluminescence. So far its use is limited to sulfur compounds. The air which is polluted with sulfur compounds is fed into a hydrogen-rich flame. The resulting $S_2$ molecule is excited and emits a characteristic band, which may be monitored through a band pass filter.

## LASERS

The methods discussed here partly overlap with other categories, but are all brought together under the glamour of the laser. The specific advantages of lasers are the low beam divergence, high beam intensity, and narrow wavelength spread. The high intensity allows an interesting variation of absorption spectroscopy. The light from a suitable laser with a wavelength corresponding to part of a strong absorption band is "chopped" by a rotating disk with slits in it. This modulated light passes into the absorption cell containing the polluted air. The pollutant absorbs the light, giving rise to periodic pressure variations which can be detected by a microphone. Because of the narrow bandwidth of the laser it is usually possible to avoid interference from other gases.

Lasers can be used to detect particulate matter. The way in which light scatters from small particles depends on the size, shape, and composition of the particle as well as on the wavelength and angle of scattering. The relationships are very complicated, but partial information about the particles present may be obtained by measuring the scattered light at many angles and with different wavelengths.

The most interesting potential use of lasers is for remote detection of a specific pollutant with positive identification of the location. One makes use of the scattering of light by molecules, often in the backwards direction. Molecules scatter strongly when light of the same energy as a transition from the ground state is incident. This is called resonant scattering. Unfortunately it is difficult to distinguish between resonant scattering and simple reflection by a particle. A variation is fluorescent scattering. Here light of one energy (wavelength) is absorbed, and light of a lower energy (longer wavelength), corresponding to a downward transition to an intermediate state, is emitted. Since the return wavelength is different and again unique, positive identification can be made. Both of these cases suffer from the general lack of availability of lasers with the proper wavelengths needed. However $SO_2$ and $NO_2$ can be detected in this way.

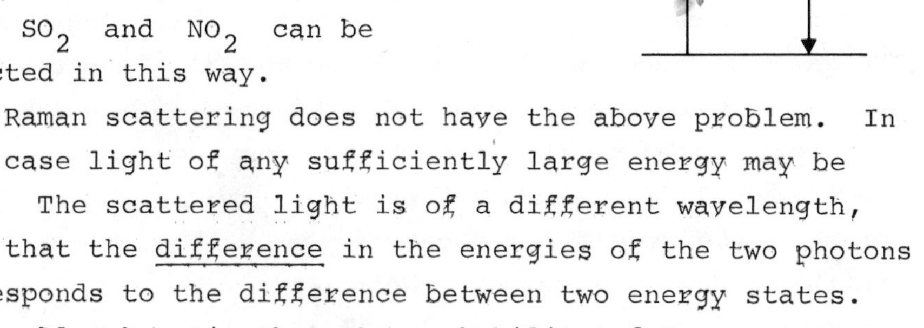

Raman scattering does not have the above problem. In this case light of any sufficiently large energy may be used. The scattered light is of a different wavelength, such that the <u>difference</u> in the energies of the two photons corresponds to the difference between two energy states. The problem here is that the probability of Raman scattering is quite small, so that it is difficult to obtain the needed sensitivity. High present expense is also a negative factor.

The great advantage of laser scattering is that distance ranging may be obtained for back-scattered light. A short pulse of light is sent out. If the time delay of the returned scattered light is measured, the distance off to the pollution may be obtained, thus pin-pointing the location.

## OTHER METHODS

The gas chromatograph is very good at separating different kinds of pollutants. It is one of the few methods which can easily separate the various hydrocarbons. A gas chromatograph basically consists of a long tube packed with some material (the kind of material depends on the gases to be detected). The gases to be studied (polluted air in our case) are introduced to the tube along with an inert carrier gas. The rate of flow through the tube depends on the gas type and the temperature. Different gases arrive at the other end of the tube at different times. Their arrival may be detected with a thermocouple gauge or a flame ionization detector. Quantitative measurements depend on a prior calibration of the instrument. The gas chromatograph can be automated with a computer to control and analyze the results. The disadvantages of this method include the fact that it must be cycled in temperature and thus can not provide a continuous monitoring, and the need for a carrier gas. Gas chromatographs are the best detectors for specific hydrocarbons. Carbon monoxide can also be detected by first chemically converting it to $CH_4$.

Mass spectrometers can be used. These directly measure the molecular mass. If kinetic energy and either momentum or velocity can be measured, mass may be determined. One form is shown in Fig. 3. The gas is ionized in the source. The acceleration region determines the energy:

$$q\ V = m\ v^2/2$$

The deflection in the magnetic field then determines the

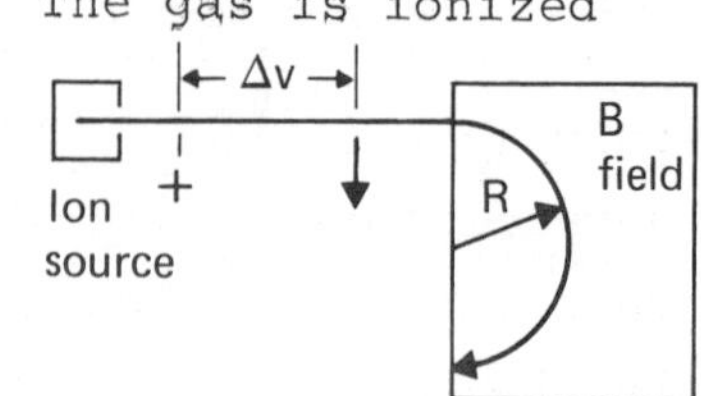


Fig 3. Schematic of one form of a mass spectrometer.

momentum

$$q\ v\ B = m\ v^2 / R$$

Elimination of $v$ between the above equations yields

$$m = q\ B^2\ R^2/(2\ V) \tag{1}$$

There are many variations on this design. Another common form uses a direct measurement of velocity from time and distance, and is called a time-of-flight mass spectrometer.

One of the problems of a mass spectrometer is that molecules are often dissociated or partially dissociated when they are ionized. Thus one often is measuring the mass of fragments, and must deduce the original molecule from the fragments. Another problem is that two different molecules may have the same or almost the same mass. A simple example is $CO$ and $N_2$. Both of their masses are about 28 (on the AMU scale). More exactly $CO$ is

$$12.0112 + 15.9994 = 28.0106$$

and $N_2$ is

$$2(14.0067) = 28.0134$$

Thus one must be able to distinguish 0.0028 out of 28, or 1 part in $10^4$. More complicated molecules also provide similar examples. In the case of some molecules, especially organic ones, two different molecules may have identical chemical formulas and thus masses and differ only in structure. Since the chemical effects may be very different, it is important to distinguish between such isomers. The existence of fragments can help identify the particular structures.

Electrochemical transducers are growing in popularity.

Basically the gas to be detected passes through a semipermeable membrane into a region where there is an electrolyte and a catalytic electrode. An electrical potential is produced. The current flowing is a function of the amount of pollutant present. The action is rather like that of a fuel cell. Interference by other pollutants and lack of long-term stability pose problems at present.

REFERENCES

1. A. Coble, _et al_., "New Eye on the Air" _Environment_, May 1971.

2. T. H. Maugh II, "Air Pollution Instrumentation: A Trend toward Physical Methods" _Science_, Aug. 25, 1972, and "Air Pollution Instrumentation (II): The Glamour of Lasers" _Science_, Sept. 22, 1972.

3. A. C. Stern, editor, _Air Pollution_, _Second Edition_ (Academic press, 1968), Vol. II.

4. J. A. Hodgeson _et al_., "Air Pollution Monitoring by Advanced Spectroscopic Techniques" _Science_ _182_, 248 (1973).

5. C. D. Hollowell and R. D. McLaughlin, "Instrumentation for Air Pollution Monitoring" _Env. Science & Tech_. _7_, 1011 (1973).

PROBLEMS

1. Estimate the rate of change in pressure resulting from the absorption of light in a nondispersive

spectrometer. Suppose that 1% of a 1 W beam is absorbed in a cell that is 1 m long and 10 cm in diameter and at atmospheric pressure.

2. If one wants to pinpoint a pollution source to the nearest 100 ft by using back-scattered light from a laser, what accuracy in timing must be used?

3. From Eqn. 1 derive an expression for $\Delta m/m$ in terms of $\Delta R$. Hint: $\Delta m = \frac{dm}{dR}\Delta R$. Using reasonable values for B and V, what $\Delta R$ does a value of $\Delta m/m = 10^{-4}$ imply?

# Unit 25
# Communications

## INTRODUCTION

Rapid long distance communication is essential to modern society. It is necessary for economic, political, and social purposes. Imagine a large corporation, with many factories, assembly plants, warehouses, and distributors trying to keep together if stage coach were the only way of communicating. Many battles have been fought after the wars had ended, back in the days when it took a month or more for the information to be carried. Now international crises can sometimes be prevented by the use of direct "hotlines". With weapons deployable to any point on the globe, it is in everybody's interest to know what is happening in all parts of the world. The communications satellites now help this.

Computers are becoming a very important part of modern society. Computers are fast enough that a major problem exists in getting information into and answers out of them sufficiently rapidly that the processing capacity can be utilized. Frequently there is a central computer connected to many remote terminals, thus increasing the information transfer problem. In some cases one computer is linked to another.

Concentrated business districts developed because of the need for people to talk to each other, back in the days when face-to-face conversation was the only way to have a prompt two-way communication. If we exploited modern technology, there would be less need for millions of office workers to leave their homes and travel to the business district every day. The transportation of communications (rather than people) does not cause traffic jams or air pollution. More recreational and educational materials could be brought into the home, further reducing the need to move people. However, a very large expense would be involved.

There are definite limitations to how much information can be transmitted over a given communications channel. We explore some of the aspects in this Unit.

## BITS AND BANDWIDTH

One of the simplest ways to transmit information is by binary digits, or "bits" as they are called. The only two digits are 1 and 0, which can be represented by "on" or "plus" and "off" or "minus", respectively. (The Morse code, with its dots and dashes, is a type of binary code.) The electronic circuitry thus need be in one of only two states, rather than ten states as with a decimal system or a continuum in the case of analog signals. If the difference between the "on" state and the "off" state is much larger than the circuit noise, there is no problem with noise.

The binary system is based on powers of 2. Starting from the right hand digit and going left, the positions correspond to $2^0$, $2^1$, $2^2$, $2^3$,.... Thus the first few binary numbers and their decimal equivalents are as shown in Table 1.

Table 1. Binary Numbers

| binary | decimal | binary | decimal |
|---|---|---|---|
| 0 | 0 | 101 | 5 |
| 1 | 1 | 110 | 6 |
| 10 | 2 | 111 | 7 |
| 11 | 3 | 1000 | 8 |
| 100 | 4 | 1001 | 9 |

One can obviously also arrange a correspondence between sets of bits and the letters of the alphabet. Thus any information which is either in words or which can be quantified can be transmitted by a binary code.

The rate at which bits can be transmitted depends on the range of frequencies carried, or bandwidth, of the channel. As the number of bits per second increases, $\Delta t$, the time duration of each obviously must decrease. A spread of frequencies, $\Delta f$, is needed to create a pulse. One way to arrive at the frequency spread is to appeal to the Heisenberg uncertainty relation for time and energy:

$$\Delta E \, \Delta t \gtrsim h$$

The energy of a photon is related to its frequency by

$$E = h f$$

Thus

$$h \, \Delta f \, \Delta t \gtrsim h$$

or

$$\Delta f \gtrsim \frac{1}{\Delta t} \tag{1}$$

As the duration $\Delta t$ becomes shorter, the spread in frequencies $\Delta f$ increases in inverse proportion. The reason the above analogy gives the proper result is that the mathematics behind the Heisenberg uncertainty relation and a Fourier analysis of a pulse of length $\Delta t$ is basically the same.

One can also obtain an approximate understanding of the time-bandwidth relation from the following example. Consider the interference of two sinusoidal waves of frequencies $f$ and $f + \Delta f$. See Fig. 1. We start the waves out of phase. Because of the frequency difference they come into phase, and then go out of phase again a time $\Delta t$ later. The lower frequency wave goes through $N$ cycles while the higher frequency wave goes through $N + 1$ cycles since there must be one extra wave for a $360°$ phase change. We thus have for the lower frequency

$$N = f\,\Delta t \tag{2}$$

and for the higher frequency

$$N + 1 = (f + \Delta f)\,\Delta t \tag{3}$$

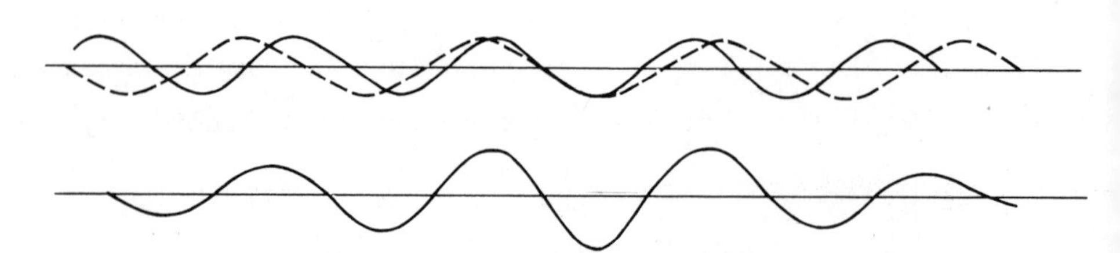

Fig. 1. The interference of two waves. The upper line shows the two waves; the bottom shows the sum.

A subtraction of Eqn. 2 from Eqn. 3 yields

$$\Delta f \quad \Delta t = 1$$

as before.

If one wants to transmit $10^6$ bits per second each pulse must be less than $10^{-6}$ sec in duration and thus have a frequency spread of about $10^6$ Hz. The communications channel must therefore have a bandwidth of around 1 MHz. A more exact relation is given later.

## MULTIPLEXING

Multiplexing means sending more than one signal over a given channel. A common example is FM-stereo broadcasting. Frequently an information source will have a limited rate of generation of information. If the channel capacity (the maximum number of bits per second which can be transmitted) exceeds the number of bits per second in a single message, several separate messages can be sent at the same time. It is cheaper and easier to increase the bandwidth of a single channel so as to allow many messages than it is to provide many individual channels.

Consider the transmission of a vocal message. The important frequencies of the voice lie between about 200 and 4000 Hz. If much less than this range of frequencies (for example, a band of less than 1000 Hz) is transmitted, the intelligibility of the message decreases. (Recall that this frequency range is important for the interference of speech by noise, as covered in Unit 10.) If a cable or carrier channel has a bandwidth of 15000 Hz it could carry three separate conversations. One way this could be accomplished is to leave the first one alone, add 5000 Hz to the second (so that the frequencies would now be between 5200 Hz and 9000 Hz), and add 1000 Hz to the third. The thre signals are then added together and transmitted. At the other end a filter tuned to 0 - 4000 Hz would separate out the first. Similarly other filters would separate out the second and third messages, which would then be frequency-shifted back down to the normal range. The gaps in frequency are needed to allow for the fact that filters are not perfect in separating different frequency regions. Clearly the bandwidth of the channel must increase in proportion to the number of separate messages. This method of multiplexing is called frequency division. A given frequency region contains only one message. If one measures the signal at some particular time, it will be composed of contributions from all of the messages.

Another multiplexing method is called time division. In this case one samples the first signal at $t_1$, the second at $t_2$, the third at $t_3$, and then back to the first signal at $t_4$, and so on (assuming three messages again). Thus at any given time the signal represents a portion of only one of the messages. However if one examines a given frequency range of the composite signal, one obtains a composite of all of the messages.

If a signal has a bandwidth of b cycles per second, it must be sampled 2b times per second to be reproduced adequately. The time-sampling of a complex signal is shown in Fig. 2. The sampling yields a series of pulses for each signal, with different sets of sampling times used for the different signals. The pulse trains are added together. Since the times of the pulses are distinct, the information of each signal remains separate. The composite signal is decoded by sorting in the proper sequence to obtain the separate pulse trains corresponding to each signal. A pulse train such as shown in Fig. 2b, contains a broad range of frequencies. If the pulse train is put through a low-pass filter with the original band width $b$ so that only the low frequency components remain, it can be shown that the original signal is recovered. Since each message is sampled $2b$ times per second, there are $2Nb$ pulses per second to be sent over the channel for $N$ messages. Thus the bandwidth of the multiplexed signal again increases with the number of messages.

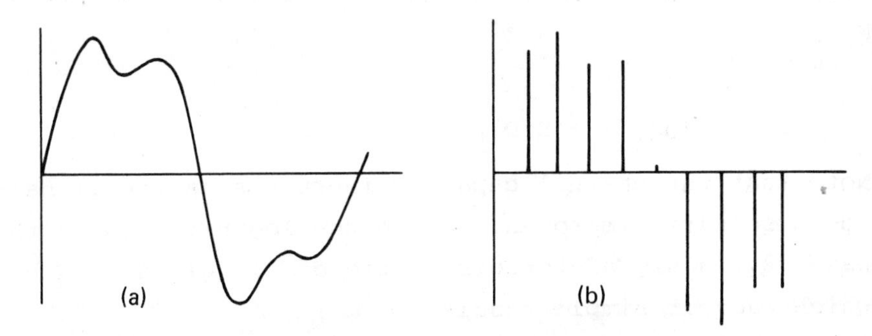


Fig. 2. A complex waveform (a), and the pulse train from a representative sampling (b).

Often a multiplexed signal is sent over a modulated carrier. In general the maximum possible bandwidth is some fraction of the carrier frequency. Thus as the carrier frequency increases, the bandwidth and hence the maximum number of messages increases. Lasers operating in the optical portion of the spectrum thus have the potential for carrying a very large number of messages, provided an optical modulator with wide bandwidth can be developed, and the laser frequency can be stabilized adequately.

## MAXIMUM INFORMATION CAPACITY

In an earlier section an approximate relation between the

number of pulses per second and bandwidth was given. The result was that the number of bits per second was about equal to the bandwidth. An electrical engineer named Claude Shannon derived a precise result for the channel capacity, or maximum number of bits per second which can be transmitted over a channel of bandwidth b. The exact relation for the channel capacity C depends on the average signal powers, S, and the average noise power, N:

$$C = b \log_2 (1 + S/N) \tag{4}$$

Note that the channel capacity increases as the signal-to-noise ratio improves. Since the logarithm is to the base 2, a signal-to-noise ratio of 1 gives $C = b$, which was our simple result.

Tables of logarithms to the base 2 are rare, so we review the method for changing base. From the definition of logarithms,

$$x = 10^{\log_{10} x} = 2^{\log_2 x}$$

If we take the logarithm (base 10) of the right-hand portion of the above equations and rearrange, we obtain

$$\log_2 x = \log_{10} x / \log_{10} 2$$
$$= 3.3219 \log_{10} x$$

As the signal-to-noise ratio gets less than one, the channel capacity decreases to less than the bandwidth. However, because of the behavior of the logarithmic function, increasing the signal-to-noise ratio beyond about 10 improves the channel capacity only slightly. See Prob. 3. While Eqn. 4 represents the maximum possible channel capacity given sufficiently sophisticated encoding and decoding devices, it is usually cheaper to settle for less than optimum channel use and provide a few more channels.

Speech does not make use of the available channel capacity; only about one third or less is used. The human voice is capable of only a limited number of the possible sounds which can be generated in the 200 - 4000 Hz range. Any language has a much smaller subset of these possible sounds. Some redundancy is built into a language so that a message can get through even in the presence of considerable noise. The redundancy of vocal messages can be shown by using a variable bandpass filter. Good intelligibility can be obtained with a band lying between 1000 Hz and 3000 Hz, and reasonable quality can be obtained with either 750 - 1500 Hz or 1500 - 3000 Hz. On the other hand if the band between 1000 Hz and 4000 Hz is eliminated, the message can still be understood! Television also makes use of only a fraction of the channel capacity. The channel is capable of transmitting an entirely different picture each scan, or 30 different pictures per second. The human eye and brain would not be able to receive or understand that much information.

## REFERENCES

1. J. R. Pierce, Waves and Messages (Doubleday, 1967). Especially Ch. 3 and 4.

2. A. H. W. Beck, Words and Waves (McGraw-Hill, 1967) Ch. 8.

## PROBLEMS

1. Convert the following decimal numbers into binary: 14, 37, 321, 1502. Convert the following binary numbers into decimal: 1011, 110100, 1010111.

2. A binary number is 16 bits long. Approximately how large is this as a decimal number? Develop a general rule for converting large binary numbers into an order-of-magnitude, and then answer the specific case given.

3. Evaluate Eqn. 4 for $b = 10^4$ Hz and S/N varying from 0.01 to $10^4$, and then draw a graph. It is suggested that the S/N abscissa be a logarithmic scale.

# Unit 26
# Nuclear Power: Fission

## INTRODUCTION

The supply of fossil fuels is definitely limited. Even though a firm inventory is not possible, reasonable estimates have been made. They show that the oil and natural gas supplies will be almost exhausted in less than one hundred years, and that the coal reserves will be depleted in a few hundred years. See Appendix 3. The time remaining is of course dependent on rates of usage and could vary somewhat from those given. There are also various environmental problems associated with burning fossil fuels. For these reasons, alternative energy sources need to be developed.

Energy can be released from nuclear reactions, either by fissioning (splitting) an appropriate heavy nucleus or by combining appropriate light nuclei. We consider the former here and the latter in Unit 28. The subject is obviously too vast to cover with any sort of completeness in two short Units. We will focus attention on the basic nuclear physics, the fuel supply, and the possible environmental impact.

## NUCLEAR BINDING ENERGY

Both fission and fusion involve the rearrangement of nucleons (the collective term and neutrons and protons) into different nuclei so that energy is released. The basis for understanding the energy release is found in the binding energy and the way it varies with the number of nucleons. The total binding energy is the amount of energy needed to disassemble a nucleus into its constituent nucleons. The main contribution to the binding energy comes from nuclear forces. Since the nuclear force is short-ranged, a nucleon interacts only with its immediate neighbors. An interior nucleon is completely surrounded and the nuclear force is said to be saturated. Its binding energy is not altered by the addition of more nucleons. The contribution to the total binding energy from interior nucleons thus increases linearly with $A$, the number of nucleons. Since the foregoing is the dominant source of binding energy, it is convenient to discuss the total binding energy divided by the number of nucleons, or the binding energy per nucleon. The contribution to this due to nuclear forces acting on interior nucleons is thus a constant, amounting to about 16 MeV per nucleon.

Not all nucleons are in the interior. Those on the surface interact with fewer nucleons and thus are bound less tightly. This surface effect reduces the total binding energy and is proportional to the surface area, which goes as the square of $r$, the nuclear radius. Nucleons behave as if they are incompressible. The nucleus therefore has approximately a constant density, and the volume is thus proportional to the number of nucleons. This leads to the following relation between radius and nucleon number:

$$r = r_o A^{1/3} \tag{1}$$

where $r_o$ has a value of about $1.2 \times 10^{-15}$ m. The surface area thus varies as $A^{2/3}$. The total binding energy is reduced in proportion to $A^{2/3}$, or the binding energy per nucleon is reduced by $- b A^{-1/3}$, where b is some constant.

The reader should carefully note that we have discussed a pictorial model of the nucleus when discussing nucleons in the "interior" and on the "surface". The actual nucleus has nucleons in motion and changing identities through charged meson exchange. The only proper description is in terms of quantum-mechanical wave functions, which are related to the probability of finding a given kind of nucleon at a given position.

So far we have been treating protons and neutrons collectively as nucleons, which is proper as far as the nuclear force is concerned. There is an important difference: the proton is charged. Since like charges repel, the Coulomb force leads to a decrease in the total binding energy. The energy associated with each pair of protons is $e^2/(4\pi\varepsilon_o r)$, or by Eqn. 1, $e^2/(4\pi\varepsilon_o r_o A^{1/3})$, where we have assumed that the protons are on the average about a nuclear radius apart. The number of pairs of protons increases approximately as $Z^2$, where Z is the number of protons, so the total binding energy is reduced by

$$- \frac{e^2 Z^2}{4\pi\varepsilon_o r_o A^{1/3}}$$

Thus the binding energy per nucleon is reduced by

$$-\frac{e^2 Z^2}{4\pi\varepsilon_o r_o A^{4/3}}$$

The numerical value of these constants, for energy expressed in MeV, is 1.2. This number should be reduced by 3/5 to take in account the spherical distribution, and if we note that $Z \simeq A/2$ (valid for the lighter nuclei), we obtain a reduction of $-0.18\ A^{2/3}$ (in MeV) to the binding energy per nucleon.

An approximate relation for the binding energy per nucleon therefore is

$$a - b\ A^{-1/3} - 0.18\ A^{2/3} \qquad (2)$$

where a and b are appropriate constants. The relative behavior of each term in expression (2) and the total is shown in Fig. 1, where we take a = 19, and b = 27. The

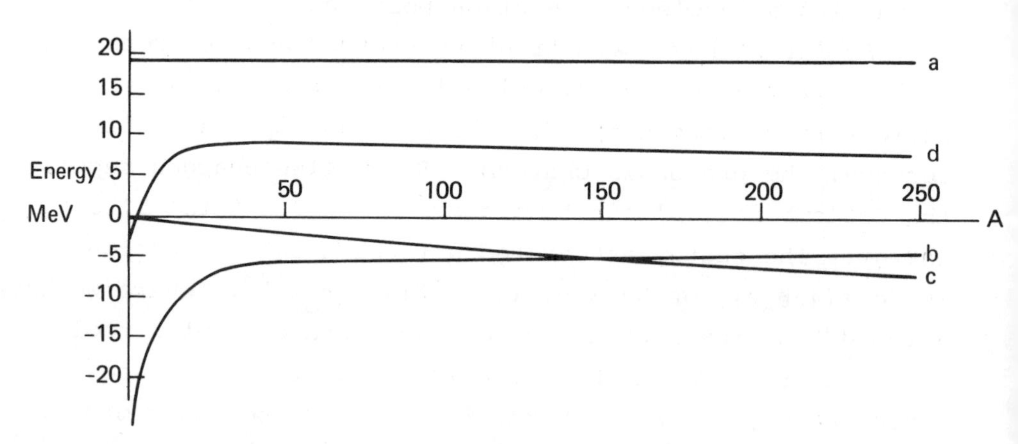


Fig. 1. Approximate contributions of various terms to the binding energy per nucleon. Line a is the constant volume term, line b is the surface correction, line c is the Coulomb energy and line d is the sum.

value $a = 19$ was used here to give a good fit to actual binding energies with only three terms in the expression. Earlier it was started that the constant term was about 16 MeV per nucleon, which is correct in a more complete treatment. The actual binding energy per nucleon is shown in Fig. 2. We note that our approximate expression is qualitatively correct. Much better models of nuclear binding energies have been made. Some of the additional terms account for the Pauli exclusion principle, shell structure, departures from spherical shape, and nuclear pairing.

The important point here is that the binding-energy-per-nucleon curve has a maximum of about 9 MeV in the $A = 50$ to $100$ range, and is lower at both ends. Energy

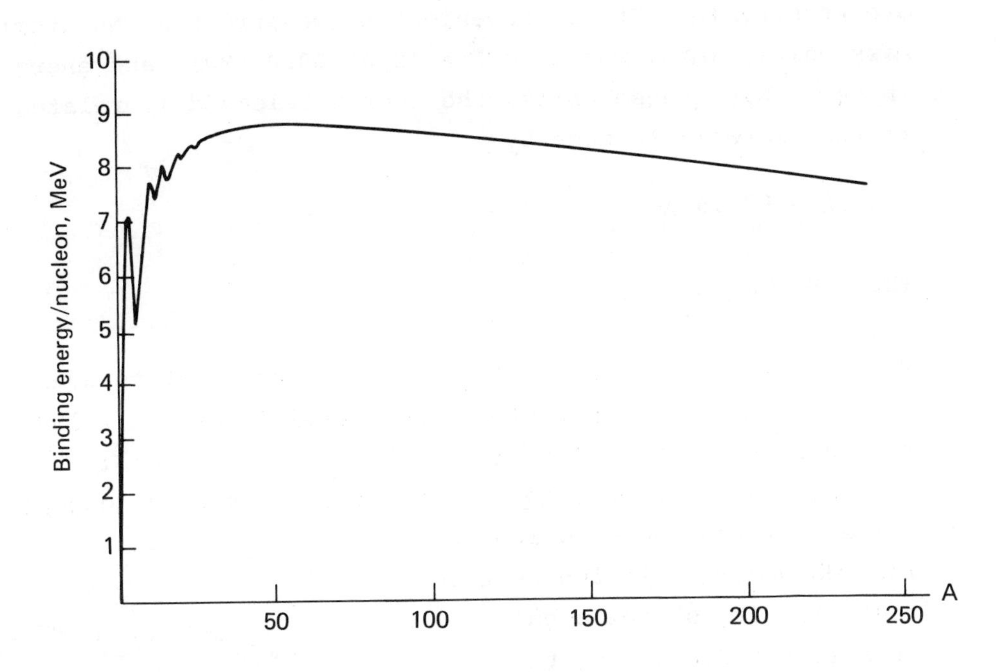


Fig. 2. Actual binding energies per nucleon as a function of the number of nucleons.

can be released if one can take a given number of nucleons and rearrange them so that the binding energy per nucleon is larger. This can be accomplished by taken a heavy nucleus ($A \simeq 235$) and splitting it into two parts or by taking two light nuclei ($A$ very small) and combining them into a single heavier nucleus. The former is fission and the latter fusion. Energies available are on the order of millions of electron volts per atom as contrasted to electron volts in the case of chemical reasctions. That is why a lot of energy can be obtained from relatively little nuclear fuel. The energies are large enough that the changes in mass required by

$$E = m c^2$$

are measurable. It is convenient to measure mass in atomic mass units (AMU, where $C^{12} = 12.0000000$ AMU) and energy in MeV. With these units, the energy released is related to the decrease in mass by

$$\Delta E = 931.5 \, \Delta m$$

## THE FISSION PROCESS

Since systems tend to go to the lowest energy state available to them, why don't all nuclei heavier than $A \simeq 150$ spontaneously undergo fission? The reason is that it takes some additional energy to get the process started (except for a very few nuclei which in fact do undergo fission by themselves). It is like a ball in a little dip at the top of a hill. It takes a push to get it started. The origin of this fission barrier can be estimated as follows. As a nucleus starts to fission, its shape becomes more and more elongated, like a football. Let us call the long axis $\alpha$ and the short axis $\beta$. As $\alpha$ increases, two competing changes in energy occur. One is associated with the Coulomb energy, which decreases since the average distance between protons has increased. The other is associated with the increase in surface area. Since a nucleon on the surface is less tightly bound than one in the interior and more nucleons will be on the surface, the energy of the system increases. The second effect is the larger one for small increases in $\alpha/\beta$. For larger values of $\alpha/\beta$ the Coulomb term becomes more important and the energy decreases. The decrease is what permits fissioning. The net result is shown in Fig. 3. This model is called the liquid drop model since it is similar to a drop splitting in two. The rise in energy to $\alpha/\beta \simeq 2$ is the fission barrier and may be different from the 6 MeV shown. Note that the minimum occurs for a slightly nonspherical shape for reasons omitted here.

For a very few nuclei (mostly in the trans-uranic region) the fission barrier is low and narrow enough that quantum mechanical barrier penetration ("tunneling") may occur, and that nucleus may undergo spontaneous fission. For other nuclei the addition of a slow neutron may provide enough energy to go over the barrier. The binding energy of a neutron is about 8 MeV in the high A region. When a neutron is captured this energy is made available. The addition of other particles, such as protons or alpha-particles (He nuclei) can also produce fissioning, but they are not considered here because they are charged particles. Thus there is Coulomb repulsion and it takes a high

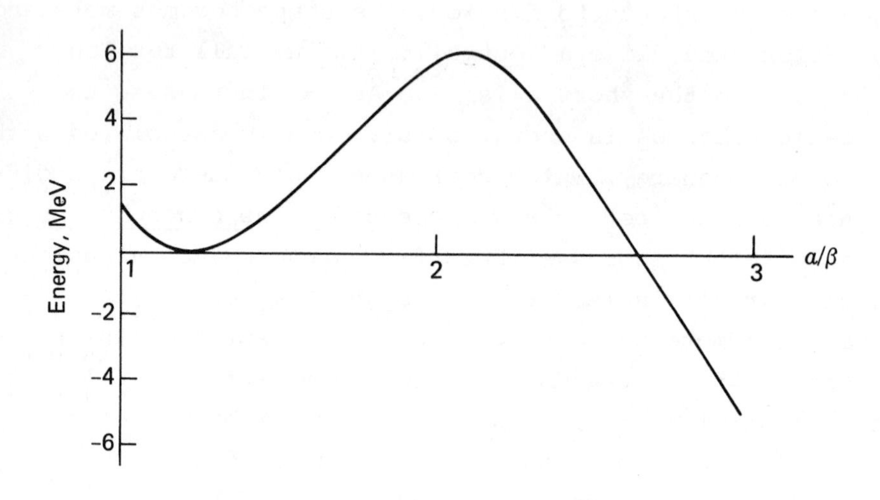


Fig. 3. The change in energy as a nucleus changes shape.

energy to get them close enough to the nucleus. The nucleus might get rid of the extra energy associated with adding the neutron by emitting a photon, or by going over the fission barrier. Only three reasonably available nuclei actually undergo fission with slow neutrons: $U^{233}$, $U^{235}$, and $Pu^{239}$, and only $U^{235}$ is found in nature. For most nuclei the binding energy of a slow neutron is not sufficient in induce fission.

We can estimate the energy released in two ways. From Fig. 2 we see that the binding energy per nucleon increases by about 1 MeV when a heavy nucleus splits into two approximately equal pieces. Since there are about 235 nucleons, we expect about a 235 MeV energy release. The actual result is somewhat less than this since the fission fragments are usually not stable ones, and hence the nucleons in them are somewhat less tightly bound. The other way is to calculate the Coulomb energy associated with the two fragments just as they separate. We suppose that one fragment is $Y^{93}$ with 39 protons and the other is $I^{135}$ with 53 protons. The Coulomb energy is

$$(39e)(53e)/[4\pi\varepsilon_o r_o(93^{1/3} + 135^{1/3})]$$

where Eqn. 1 was used for the radii. An evaluation of the above expression gives 256 MeV.

When a nucleus undergoes fission the probability is greatest that one fragment will be a little more massive than half and the other a little less massive than half. One of the many possible splits when $U^{235}$ fissions is

$$U^{235} + n \rightarrow Ba^{141} + Kr^{92} + 3n$$

The reader should verify that charge and nucleon number are conserved in this reaction. A very important feature is that some free neutrons are emitted. On the average there are about 2.5 neutrons per fission of $U^{235}$. This is one of the keys to a chain reaction. One neutron will start it, and then the neutrons released can keep it going. Since 2.5 neutrons are released, a rapidly increasing reaction rate would result if all were utilized for new reactions. To sustain a reaction at constant level, exactly one new reaction must result for every reaction which has taken place. Slightly less and the level drops; more and it rises. Neutrons are lost out the outer surface of a reactor, so the amount of fissionable material and the geometry are important. Some neutrons are captured by other materials in the reactor. By careful design a rate near one can be achieved. Final "tuning" is provided by control rods which readily absorb neutrons and which can be inserted more or less to achieve steady running.

Only $U^{235}$ of the fissionable nuclei is found naturally, and this isotope is only about 0.7% of the natural uranium. The plentiful $U^{238}$ can capture neutrons without fissioning, making it very difficult to maintain a reaction. Most power reactors use enriched uranium, where the percentage of $U^{235}$ has been increased to around 3%. This may be accomplished by the diffusion of gaseous $UF_6$ through porous barriers. Since $U^{235}F_6$ is slightly lighter than $U^{238}F_6$, it has a higher average speed and goes through slightly faster. The average kinetic energies are the same, so the average speed varies inversely as the square root of the masses. The speed ratio is thus $(352/349)^{1/2} = 1.0043$. Many stages of enrichment are therefore needed. A gaseous centrifuge may also be used for enrichment. This process is in use in Europe and is being tested in this country.

Another factor which increases the probability of $U^{235}$ fissioning instead of a neutron being captured by $U^{238}$ is the use of slow neutrons. The probability of $U^{235}$ fissioning with a slow neutron is very large compared to the probability of $U^{238}$ capturing a slow neutron. The neutrons given off by the fission reaction have energies of several MeV and hence are fast. They can be slowed down by elastic collisions on light nuclei, such as hydrogen or carbon. The material which slows the neutrons down is called a moderator. We leave further details of reactor design to the References.

## FUEL SUPPLY AND BREEDING

There is a large uncertainty in estimating the uranium supply. It is a trace element widely found in very low concentrations. The problem concerns the fact that only 0.7% of uranium is $U^{235}$. If only the $U^{235}$ is used, it is economical to mine and process only the relatively rare high grade ores. The extent of these ores is not well known, since relatively little prospecting has been done in recent years. The A.E.C. estimates that the high grade reserves may be used up in 25 or 30 years at present growth rates for nuclear energy. Even if this under-estimates the supply by a factor of 2 to 4, the high grade deposits are limited and will not last long.

There is a way to utilize the much more abundant $U^{238}$. While $U^{238}$ will not fission with a slow neutron, it can be turned into something which will. The process (called breeding) starts with the capture of a neutron and is followed by two beta decays:

$$U^{238} + n \rightarrow U^{239}$$

$$U^{239} \rightarrow Np^{239} + \beta^- + \bar{\nu} \quad (2.3 \text{ min})$$

$$Np^{239} \rightarrow Pu^{239} + \beta^- + \bar{\nu} \quad (2.3 \text{ days})$$

Plutonium-239 will fission with a slow neutron. Plutonium-239 is also radioactive, but the half-life is 24400 years, so little decays before it can be utilized. A similar capture-and-decay process converts $Th^{232}$ into $U^{233}$, which also fissions readily.

The fission process releases about 2.5 neutrons per reaction on the average. One of these is needed to sustain the reaction, leaving 1.5 on the average for the above breeding reactions. It is thus possible to create more fissionable material than is consumed. It is estimated that the supply of fissionable material could be doubled in ten years for the $U^{238}$ - $Pu^{239}$ cycle and in 20 years for the $Th^{232}$ - $U^{233}$ cycle. The former doubling time coincidentally is the same as the present doubling time for the use of electricity. (See Appendices 1 and 3.) We do not discuss breeder reactor design here; see the References.

With the full utilization of uranium which breeding allows, it becomes worthwhile to go after much lower grade ores. An an example consider the Chattanooga Shale which underlies a considerable portion of the Midwest of the U.S.A. The uranium is only 60 parts per million by weight. Yet it can be mined and processed economically with breeding. The energy released by fission is so large that a 20 mile by 20 mile area of this low grade ore has as much energy as the total of all U.S. fossil fuels. Low to medium grade thorium deposits are found in many places. At the present it is impossible to estimate the full extent of the low grade uranium and thorium deposits which could be used by breeding. Preliminary estimates are in the range of 10 000 to 1 000 000 years, assuming some leveling off of energy usage at a reasonable rate.

## SAFETY AND HAZARDS

The great controversy about fission energy centers on safety. It is a subject laden with emotions on both sides. We will not attempt to cover all issues here but merely point out the major potential hazards. First of all it should be noted that a power reactor can not explode in the manner of a nuclear bomb. The dangers are largely connected with the fact that most of the fission fragments are radioactive. The effects of radiation (primarily beta and gamma rays in this case) are covered in the next Unit, but briefly large doses kill and smaller doses can cause cancer and genetic damage. There is no general agreement as to whether some small amount of radiation is all right or whether there is no such thing as a "safe level" for radiation.

The amount of radioactive material produced in a single year in just one large nuclear power plant is equivalent to that produced by about 20 megatons of bombs. There are several paths which <u>might</u> let some or all of this radioactive material enter the biosphere. The nuclear power proponents claim that there are enough safeguards that these possibilities will not actually happen. The opponents are afraid one of them may happen accidentally.

The first possible route is from the reactor core to the outside. A fuel rod can crack letting radioactive material into the coolant. A leak could develop in a coolant line. Since there are several layers of barriers, it is not likely that a leak would reach the outside. However, unlikely things sometimes happen. Radioactive gaseous fission fragments (Kr and Xe) are released at low levels to the atmosphere. The most worrisome possibility is an accident, such as the loss of the coolant, which allows the reactor core (the fuel rod assembly) to melt down. There is sufficient heat generated by the decay of the fission fragments that the molten core would melt through any containment shield or barrier and down into the earth, thus releasing essentially all of the built-up radiation. Reactors are designed with a back-up emergency core cooling system. Unfortunately a back-up system has never had a full scale test, so it is not clear that it will function as desired. For years the A.E.C. has maintained that a full melt-down of the core simply could not occur. Yet accidents beyond what was believed possible have occurred.

Another way radioactive material could reach the environment is during transportation or processing of the used fuel elements. After a period of time the fuel rods must be treated to remove the fission fragments. Otherwise the fragments absorb too many neutrons. Since there is still unused fuel, the rods must be processed rather than just disposed. No transportation system can eliminate accidents. The fuel rods are carried in special containers that should contain the radiation even in the case of an accident. Theft of the fissionable material, especially plutonium, which can be chemically separated into bomb-grade material, remains a problem during transportation. The processing plant must handle large quantities of radioactive material and move it from one place to another. In spite of many precautions a spill could happen. It has been proposed to put many power plants and the processing plant together in some remote area to reduce some of the above hazards.

The outstanding unsolved problem is the storage of the radioactive waste. It is thermally hot from the radiation energy and often quite corrosive. The storage tanks now used last only about 20 years. Some have started leaking while still containing radioactive waste. This material must be isolated from the environment for periods of time ranging from hundreds to millions of years before decay has rendered them harmless. The isotopes which cause the most concern are those which are produced plentifully, can be incorporated into biological material, and have intermediate half-lives (on the order of 1 to 100 years). Isotopes with short half-lives can easily be stored until they decay.

Those with very long half-lives have low rates of decay. Some of the major concern is thus with $Sr^{90}$ and $Cs^{137}$, both with half-lives around 30 years. As an example, the amount of $Sr^{90}$ from a single reactor may be $10^{12}$ times the "safe" level. The half-life of $Sr^{90}$ is 28 years. To reduce the activity by $10^{12}$ takes 40 half-lives ($2^{40} = 1.0995 \times 10^{12}$), or 1120 years. Mankind has no experience with the durability of materials under the action of radiation for time spans that long. The continuation of care for generations hence, with possible changes in political and social structures, is awesome. Salt mines and the ice field of Antartica have been suggested. There are difficulties and doubts about both. As nuclear reactors continue to run and more reactors are built, the amount of material to be contained will grow larger for a long, long time. Review Unit 18.

No power generation is without its hazards and social risks. One must weigh alternatives. For example, how do the deaths from air pollution associated with burning coal compare to the deaths from radiation? At the present far more deaths are caused by the sulfur in coal than by radiation from reactors. This might change in the future because of storage problems and accidents. The total risk-benefit picture must be assessed as best as possible, with all alternatives considered. The answer is not easy. All elements of society must be consulted. Fortunately there exist other long-term energy sources with less risks.

REFERENCES

1. D. R. Inglis, Nuclear Energy (Addison-Wesley, 1973).

2. S. Novick, The Careless Atom (Delta, 1969).

3. G. T. Seaborg and J. L. Bloom, "Fast Breeder Reactors" Scientific American, Nov. 1970.

4. G. T. Seaborg and W. R. Corliss, Man and Atom (Dutton, 1971).

PROBLEMS

1. Suppose that instead of exactly one new reaction for every one occurring, thus sustaining the chain reaction, there are 1.0005 new reactions. If the time between reactions is $10^{-4}$ sec, how much will the reaction rate increase in 1 sec? This is why control methods are touchy.

2. Verify the coefficient of the Coulomb term ( $- 0.18 A^{2/3}$) in expression (2).

3. Verify that 1 AMU is equivalent to 931.5 MeV.

4. Find the actual energy released in the following fission

$$U^{235} + n \quad Ba^{141} + Kr^{92} + 3n$$

by finding the change in mass and converting to energy units. The masses in AMU are $U^{235}$ = 235.0409, n = 1.0086, $Ba^{141}$ = 140.9141, and $Kr^{92}$ = 91.9250.

5. Find the energy content in the Chattanooga Shale per square mile of surface area. The uranium is 60 parts per million by weight. Assume the shale

is 15 ft thick, the density of the rock is 2.5 gm/ $cm^3$, and that

a) only the $U^{235}$ is used, or

b) that all of the uranium is used by the breeding process.

6. How much $Kr^{85}$ is produced in one year by a nuclear power plant with an electrical output of 1000 MW? Check the overall efficiency from Unit 14 to find the input power. Krypton-85 has a 10.6 year half-life, and is produced in 0.3% of the reactions. Assume the plant operates at full capacity, and neglect the small amount of $Kr^{85}$ which decays during the year. What is the activity in Curies of this much $Kr^{85}$? One Curie is a decay rate of $3.7x10^{10}$ per second.

# Unit 27
# Effects of Radiation

## INTRODUCTION

Some awareness of "radiation sickness" came soon after the discovery of radioactive decay. However a good understanding of the effects was not obtained until much later. General public knowledge of the adverse effects of radiation came only after the extensive atmospheric testing of nuclear weapons. There has been a gradual diminishing of the fall-out from weapons following the agreement to stop above-ground testing by the U.S.S.R. and the U.S.A. The increasing number of nuclear power plants has brought concern over radiation back again.

The main types of radiation are $\alpha$ particles (a composite of two protons and two neutrons, which is the same as the helium nucleus), $\beta$ particles (electrons and positrons), and $\gamma$ and x-rays (high energy photons). The principal sources are radioactive nuclei, x-ray machines, and cosmic rays. Some radiation has always been present.

Certain of the radioactive nuclei have sufficiently long half-lives that they and their decay products are still around. These and the cosmic rays make up the unavoidable natural background radiation. So the question is not one of avoiding radiation altogether, but rather one of not adding significantly to the existing radiation levels. The catch word in the previous sentence is "significant". At the present there is no certainty about the long term effects of small additions to the radiation. Thus some people feel that a certain additional amount is not significant, while others feel that it is. The use of x-rays for medical diagnostics and treatment does not hit all individuals equally but when averaged over the population it adds appreciably to the natural radiation levels.

The major concern right now is that man might add significant quantities of radiation by using nuclear reactors. The fissionable fuel is an $\alpha$ emitter. The fission products are mostly radioactive. As one goes to the heavier end of the table of elements, the number of neutrons gets progressively larger than the number of protons. When a heavier nucleus splits, the fragments thus usually have an excess of neutrons when compared to stable elements with the same number of protons. They are therefore $\beta^-$ emitters, and, because the nuclei may be left in excited states, $\gamma$ emitters. Fission fragments can enter the environment from an accident at the reactor or during transportation, from the fuel rod processing plant, or from leaks where the waste products are stored.

## UNITS OF MEASURE

There are a variety of units in common use for measuring radiation and its effects. The first one we consider is the Curie (abreviation: Ci), which is a measure of the number of nuclear decays taking place per second. The size of the unit was originally set to be equal to the decay rate of 1 gm of radium ($Ra^{226}$). It is now defined as $3.7x10^{10}$ decays per second. The decay rate is related to the number of nuclei present by (see Unit 18, Eqns. 1 and 3):

$$\frac{dN}{dt} = - \lambda N$$

$$= - \frac{.693}{T_{1/2}} N$$

where $\lambda$ is the probability of decay per second, and $T_{1/2}$ is the half-life. Thus the activity in Curies for a given number of nuclei is high for a short half-life and low for a long half-life. See Prob. 1. Some of the radiation standards are expressed in terms of a maximum amount in Curies of a certain radionuclide which can be in the body.

The other main units quantify the effect of the radiation emitted. There are several which are almost equivalent, thus leading to some confusion. Historically one had the Roentgen, which was defined in terms of the ionization produced by x-rays and $\gamma$ rays. One Reontgen of radiation will create $3.336x10^{-10}$ Coulombs of positive and of negative ions in each cubic centimeter of dry air. (The number is 1 unit of charge in the cgs (esu) system, and implies over $2x10^9$ charged pairs is created per $cm^3$.) A newer unit, the rad, has just about replaced the Roentgen. The rad is defined in terms of the energy absorbed by matter due to the passage of radiation. A dosage of 1 rad means that $10^{-2}$ J were absorbed by each kilogram of matter. In dry air 1 Roentgen of $\gamma$ rays deposits $.88x10^{-2}$ J/kg, so the rad and the Roentgen are close to each other in air, and also, it turns out, in biological tissue. The damage to cells due to a dosage of 1 rad varies with the kind of radiation, as we shall discuss below. It has been found convenient to introduce still another unit to take this into account: the rem (rad-equivalent-man). This unit takes the exposure in rads and multiplies it by a factor (called the relative biological effectiveness, or RBE) which varies from 1 for x-rays and low energy $\gamma$-rays to about 10 for $\alpha$ particles. Frequently radiation standards are in terms of $\gamma$ radiation, in which case rads and rems are the same.

## INTERACTION OF RADIATION WITH MATTER

Charged particles cause ionization continuously along their paths. The Coulomb force interacts with the outer electrons in the atoms and in some cases imparts sufficient energy to remove the electrons completely, thus making an ion pair. In other cases the atom may be raised to a higher energy state without ionization. We can estimate the energy lost by charged particles from the data given previously: 1 R deposits $8.8 \times 10^{-3}$ J/kg and creates $3.336 \times 10^{-10}$ C/cm$^3$ of ions of each sign. The number of ion pairs is $3.336 \times 10^{-10}$ C divided by the charge of the electron, $1.602 \times 10^{-19}$ C, or $2.08 \times 10^9$ ion pairs. This number is in 1 cm$^3$ of dry air with a mass of $1.292 \times 10^{-6}$ kg. Thus in a kilogram we have $2.08 \times 10^9 / 1.292 \times 10^{-6} = 1.611 \times 10^{15}$ ions. The energy lost per ion pair is thus $8.8 \times 10^{-3}$ J$/1.611 \times 10^{15} = 5.46 \times 10^{-18}$ J, or 34 eV. Actual experimental values range from 22 eV to 42 eV, depending on the material.

Because each ion pair takes a certain amount of energy and the ion pairs are created with reasonable regularity along the path, the distance that a charged particle with a given energy will go is well defined. This distance is called the range. It obviously depends on energy, increasing very approximately as the energy raised to a power lying between 1 and 1.75. The range for a given energy depends of the kind of particle. First of all an $\alpha$ is doubly charged, so the Coulomb force is twice as great at a given distance as for an electron or positron. Secondly an $\alpha$ particle has about 8000 times the mass of an electron, so for a given energy it is moving much more slowly (see Prob. 2). Thus it spends more time near a given atom and the impulse is greater, leading to a greater transfer of momentum. The range of a 1 MeV $\alpha$ in air is about 0.5 cm, while an electron of the same energy will go about 400 cm. In soft living tissue the ranges are $7 \times 10^{-4}$ cm and 0.43 cm, respectively, for the same energy. There is some small variation in the range, called straggling, due to the statistical nature of the interactions.

It is instructive to find what fraction of the molecules are ionized by charged particles in water. We use the 1 MeV figures given above, since the range in water is about the same as in living tissue. One cubic centimeter of water has a mass of 1 gm. The molecular weight of water is 18, so each 1 $cm^3$ has $3.34 \times 10^{22}$ molecules. The average distance between molecules is $(3.34 \times 10^{22})^{-1/3} = 3.11 \times 10^{-8}$ cm. The range of a 1 MeV $\alpha$ is $7 \times 10^{-4}$ cm, so it passes directly by $2.25 \times 10^{4}$ molecules. Since about 33 eV are lost per ion pair, a 1 MeV particle produces about $3 \times 10^{4}$ ion pairs. Thus slightly more than one atom per water molecule is ionized by a 1 MeV $\alpha$. An electron has about 600 times the range, so only about one atom per 450 water molecules is ionized. Biological tissue has about the same density as water. Because most of the molecules are much larger than water, we must put things on a per-atom basis for comparison. About every other atom is ionized by a 1 MeV $\alpha$, and only about one in a thousand is ionized by a 1 MeV electron. This is the basis for the modifying factor for $\alpha$ particles in the rem. In air a 1 MeV $\alpha$ will ionize about one atom out of 100. See Problem 3.

The interaction of x-rays and $\gamma$ rays is quite different. Since a photon is an uncharged particle, it does not produce ionization continously along its path. However, since it is the quantum of the electromagnetic field, it can interact with charged particles. The probability of the interaction is sufficiently low that the statistical nature is macroscopically evident. One can assign a probability of interaction per unit length, $\mu$. The attenuation in intensity is then

$$dI = -\mu I \, dx$$

which becomes

$$\frac{dI}{I} = -\mu \, dx$$

The integration of the above yields

$$I = I_o e^{-\mu x} \tag{1}$$

The absorption coefficient $\mu$ is often given divided by density, and thus in units $cm^2/gm$. Equation 1 shows that the intensity of x-rays and $\gamma$ rays falls off exponentially with increasing distance. So rather than having a well-defined range as in the case of charged particles, one has a mean or $1/e$ distance. The mean distances for a 1 MeV $\gamma$ are about 100 m for air, 16 cm for water, and 1 cm for lead.

The absorption coefficient is a complicated function of photon energy and the atomic number (Z) of the absorber.

This is in part because there are three different processes by which photons interact. The first, which dominates at energies below about 300 keV, is the photoelectric effect. In this process the photon is completely absorbed, its energy going to knock an electron out of an atom. In general $\mu_{photoelectric}$ gets smaller as energy increases. The exceptions occur when the photon energy just becomes large enough to remove an electron from one of the inner shells. At moderate energies (up to about 5 Mev) Compton scattering is the most important process. In this case a photon scatters from an essentially free electron. Energy and linear momentum are conserved, so the scattered photon is of lower energy. The important high energy interaction is pair production. Here a positron-electron pair are created from the energy of the photon. This process must take place near a charged particle so that momentum can be conserved. See Problem 4. Since the rest energies of a positron and electron sum to about 1.022 MeV, it is clear that this is the lowest photon energy for pair production. The absorption coefficient generally rises with energy for pair production. Ultimately the positron comes to rest and then annihilates with an electron. Two 0.511 MeV $\gamma$ rays result, which can interact further.

The effects of $\gamma$ rays on matter are mainly second-hand, coming from the ionization produced by the electrons set in motion by the above processes.

Because $\alpha$ and $\beta$ particles have such short ranges, they are a health hazard only when the emitting nucleus is inside the body. However it is rather easy to inhale or ingest radio-active material. Gamma rays can interact within the body even when the source is external.

## EFFECTS OF RADIATION ON LIVING MATTER

It should be clear from the above discussion that the effects are due to the ionization produced. Effects can be broadly divided into somatic (damage to the organism exposed) and genetic (damage to future generations). There is some fuzziness in this division, since damage to the code whereby cells reproduce can also affect the organism which was exposed.

If a sufficient number of ionizations (on the order of $10^6$ to $10^7$) takes place in a single cell, the cell will be killed. Since the density of ions along the path of an $\alpha$ particle is much greater than for an electron, fewer $\alpha$ particles traversing a given cell are needed to kill a single cell. On the other hand a given electron will pass through far more cells because of its greater range, so enough electrons to kill a single cell will result in many neighboring cells being affected as well. Single acute doses of radiation start to be fatal to man when the whole body receives around 300 rem. About 50% of the people exposed will die at a whole body dosage of 500 rem, while 1000 rem is almost always fatal. Because repair mechanisms exist in the body the lethalness of a given dose decreases as it is spread out over longer and longer periods of time. Extremely large doses delivered to only a part of the body, such as an arm or a single kidney will not result in fatality. Once a cell is dead, further radiation does not matter. It would take an extra-ordinary reactor accident to expose more than a few people to acute lethal doses.

The more subtle problem arises when a cell is only partially damaged. It may still live, but not function properly. If the chromosomes which govern the reproduction of a cell are damaged, the cells may either not reproduce at all, leading to a premature aging effect, or may reproduce defectively. The chromosomes may either be damaged directly by the radiation, or indirectly. The indirect effects are caused by the chemical action of free radicals which are produced by the radiation. Often the indirect effect is the more important one. If the growth rate of the defective cells is uncontrolled, one has a cancer. (Leukemia is included under the general heading of cancer for convenience here.) Even very small doses of radiation could cause a cancer, while very large doses might not. This is because it is a random change whether the chromosomes will be affected by the radiation, and further random chance whether the affected chromosomes will rearrange in just such a manner as to cause a cancerous reproduction. One particle might be enough or billions might not. Significantly, cancer does not usually become obvious until 10 to 20 years after the exposure to the radiation.

The recent debate and controversy over the effects of low level radiation have centered on the cancer problem. Some argue that there is no such thing as a "safe" level and that every little bit has its effects. This is called the linear theory. So far experiments on animals and statistical analysis of people exposed by weapons or accidents have been limited to larger doses and thus have not been able to settle the issue. The argument in the preceding paragraph favors the linear theory. Certainly until it is disproven, the conservative thing to do is assume the linear theory correct. It has been shown that 100 rad approximately increases the naturally occurring cancer rate by 100%, but the figure varies somewhat with the particular kind of cancer. The linear theory would imply that 1 rad would increase cancer by 1%, and so on. Gofman and Tamplin estimated if the entire population of the U.S. were to be exposed to 0.17 rad/yr that there would be about 30000 additional cases of cancer per year. A recent report by the National Academy of Sciences places the number in the range 3000 - 15000. In 1971 the radiation limits were revised. Previously the A.E.C. limit was that the maximum additional dose to an individual (non-nuclear worker) was 0.5 rem/yr and to the population as a whole, 0.17 rem/yr. The new standards now put the limit at 0.005 rem/yr for radiation from water cooled nuclear power plants. The new philosophy is to set the standards as low as technology will allow. Previously the standard was sort of a license to pollute up to a limit.

If the genetic code of a sex cell is altered, genetic damage (mutations) will show up in succeeding generations. The mutations caused by radiation are often recessive, so the fact that there was no apparent increase in mutations in children born to survivors of Hiroshima and Nagasaki is not conclusive. All of the firm quantitative information on radiation-induced mutations is on plants, insects, and small laboratory animals. These results are difficult to extrapolate to humans. Again it matters whether a given dose is received in a short period of time or is spread out over a long time. The National Academy of Sciences has estimated that 0.17 rem/yr to the entire U.S. population would cause perhaps 500 to 10000 serious mutational defects and up to 27000 less serious ones per year after several generations.

## NATURAL RADIATION

One of the arguments for permitting certain levels of man-produced radiation is that nature bombards man with considerable radiation all of the time. Some of this radiation comes from naturally occurring isotopes such as $Ra^{226}$, $U^{238}$, $Th^{232}$, $K^{40}$, $H^3$, and $C^{14}$. Potassium-40 is found everywhere that potassium is, including in the human body. Tritium ($H^3$) and $C^{14}$ are continuously produced and generally incorporated in living matter. The radiation level received by an individual from these sources varies considerably, depending on the mineralogy of the area and even the type of house a person lives in. The range may be from 10 to 120 millirem/yr from radia-active sources external to the body. About another 20 mrem/yr comes from radioisotopes within the body. Cosmic rays supply a significant additional dosage. Again there is considerable variation. First there is an altitude effect, since the earth's atmosphere absorbs some of the radiation. At sea level the dosage is around 30 - 40 mrem/yr. This increases to 60 - 80 mrem/yr at 5000 ft and to 100 - 160 mrem/yr at 10000 ft above sea level. There is also a variation with lattitude, since the primary cosmic rays are mostly charged particles. At the magnetic poles the motion of the particles is more nearly parallel to the direction of the magnetic field, so there is little magnetic force. At the equator, the particles are traveling more nearly perpendicular to the magnetic field, and the force is large enough that the lower energy ones are returned to space. When averaged over the entire U. S. population the average natural exposure is around 90 mrem/yr. Medical uses of radiation, when averaged over the population as a whole, add about 50 - 100 mrem/yr. Fall-out from bomb testing still amounts to about 5 mrem/yr. Thus the total radiation received at present averages about 170 mrem/yr. It is not clear what fraction of the existing cancer and birth defects is caused by the unavoidable natural radiation.

Since there is no obvious difference in cancer and other deaths for people exposed to the various different levels of natural radiation (excluding small groups exposed to very high levels due to unusual mineralogy), there was a general feeling that some additional radiation above the natural would not be detrimental. Apparently the old 0.17 rem/yr standard was set on the assumption that the existing level of radiation could be doubled. There is no obvious reason why a 100% rather than 50% or 200% increase was adopted. The new outlook is "the less radiation the better".

Another factor which makes the setting and enforcing of standards difficult is the concentration of radioisotopes in the food chain. While levels of radioisotopes emitted into the biosphere may be within allowed limits, they can be concentrated to dangerous levels. Plants draw nourishment from the surrounding ground. Certain elements may be incorporated into their structure and concentrated. A grazing animal may then eat many of these plants, and then man may eat the animal, providing further and still further concentration. This increase has been well documented in ponds near Oak Ridge, Tennessee. The average concentration of $Cs^{137}$ in the water was $3.3 \times 10^{-14}$ Ci/gm (Curies/gm), while in fish it was $3.5 \times 10^{-11}$ Ci/gm. The concentration of other isotopes showed levels 2000 to nearly 9000 times higher in fish than in the water. Radioisotopes of elements which are incorporated in living systems, such as iodine, or which are chemically similar to ones which are, such as cesium (similar to potassium) and strontium (similar to calcuim) are particularly subjec to concentration.

REFERENCES

1. J. Harte and R. H. Socolow, Patient Earth (Holt, Rinehart, and Winston, 1971) Ch. 18.

2. D. R. Inglis, Nuclear Energy (Addison-Wesley, 1973) Ch. 5.

3. L. Hodges, Environmental Pollution (Holt, Rinehart, and Winston, 1973) Ch. 15.

4. J. A. Lieberman, "Ionizing-radiation Standards for Population Exposure" Physics Today, Nov. 1971.

5. The Beir Report, "Effects on Populations of Exposure to Low Levels of Ionizing Radiation" Bull. Atomic Scientists, Mar. 1973.

6. "Natural Radiation" Environment, Dec. 1973.

PROBLEMS

1. Find the activity in Curies of $10^{16}$ atoms if the half life is a) two hours, and b) 100 years. Also find the activity of each four hours and two months later.

2. Find the velocities of a 1 MeV electron and a 1 MeV $\alpha$ particle. Note that relativistic equations must be used for the electron.

3. Show that about one in every 50 molecules is ionized by a 1 MeV $\alpha$ particle in air. Why is this number so much smaller than in water?

4. Show that energy and momentum can not both be conserved in the production of a positron-electron pair by a photon in free space. This is why it occurs in some electromagnetic field. Use relativistic mechanics.

5. Calculate the force and radius of curvature of a cosmic ray proton in the earth's magnetic field (assume 1 Gauss) if the energy is 1 MeV, 1 GeV, and 100 GeV.

6. How much energy is deposited in an adult body when there is a whole-body dose of 100 rad?

# Unit 28
# Nuclear Power: Fusion

## INTRODUCTION

Solar energy is the only proven very long term major energy source. Energy from nuclear fusion has the promise of being another source of long duration. Controlled fusion has not yet been demonstrated in the laboratory, but researchers are approaching the necessary conditions. Even after feasibility is established, there will be very difficult engineering problems to be solved before commercial power plants are made. It is reasonable to suppose that significant nuclear fusion power is at least 50 to 75 years off. However, the large fuel supply and the relatively minor environmental impact make fusion appear to many to be the most attractive alternative for our energy needs.

We saw in Unit 26 that the binding energy per nucleon curve had a maximum in the $A = 50 - 100$ region. Since energy is released whenever nucleons can be rearranged into new nuclei with higher binding energy per nucleon, we can obtain energy by combining or fusing light nuclei together.

Fusion is the basic source of energy of stars. In the "main sequence" state, which is the principle one, four hydrogens are put together to form a helium nucleus. The smaller stars use the following reactions:

$$H^1 + H^1 \rightarrow H^2 + \beta^+ + \nu \quad (1a)$$

$$H^2 + H^1 \rightarrow He^3 + \gamma \quad (1b)$$

$$He^3 + He^3 \rightarrow He^4 + 2H^1 \quad (1c)$$

There is also an additional set of reactions involving $He^3$ and $He^4$ which replaces Eqn. 1c with somewhat less probability of occurring. The more massive stars (more than about 2 or 3 time the mass of the sun) turn hydrogen into helium by another set of reactions called the CNO cycle. In this cycle $C^{12}$ acts as a catalyst. The first reaction in the set for lighter stars, Eqn. 1a, has an extremely small chance of happening, since it involves the "weak" interaction. It can occur in stars since the temperature, density, and volume are all large enough that the improbable happens. The main steps of the CNO cycle involve the reaction of hydrogen with nuclei having 6 or more protons. There is thus considerable Coulomb repulsion, and therefore very high temperatures are needed. This is why the CNO cycle occurs only in massive stars.

Fortunately there are some other fusion reactions which are practical on earth. We discuss these below. It is assumed here that the reader is familiar with Units 20 and 26.

## BASIC REACTIONS

Hydrogen has two heavier isotopes: $H^2$ or deuterium (symbol: d) and $H^3$ or tritium (symbol: t). The practical fusion reactions mostly involve these isotopes. The main reactions are:

$$H^2 + H^2 \rightarrow H^3 + H^1 + 4.03 \text{ MeV} \quad (2a)$$

$$\text{or } H^2 + H^2 \rightarrow He^3 + n + 3.27 \text{ MeV} \quad (2b)$$

$$H^2 + H^3 \rightarrow He^4 + n + 17.59 \text{ MeV} \quad (2c)$$

$$H^2 + He^3 \rightarrow He^4 + H^1 + 18.35 \text{ MeV} \quad (2d)$$

The first two alternative final products (Eqns. 2a and 2b) occur with about the same probability. The last reaction, Eqn. 2d, involves the reaction of a single charge with a double charge, and therefore needs a higher temperature than the others, which only involve single charges.

All of the reactions, Eqn. 2, release energy. The amount of energy is most easily found by comparing the mass afterwards to the mass before. The difference appears in the form of energy, at the rate of 931.5 MeV/AMU. The masses for Eqn. 2a are

$H^1$ 1.007825 AMU

$H^2$ 2.014102

$H^3$ 3.016050

The masses before and after are

| | | | |
|---|---|---|---|
| $2\ H^2$ | 4.028204 | $H^1$ | 1.007825 |
| | | $H^3$ | 3.016050 |
| | | | 4.023875 |

The mass afterwards is smaller by 0.004329 AMU, which

amounts to 4.032 MeV. Similarly the energy calculation for Eqn. 2c is

| | | | |
|---|---|---|---|
| $H^2$ | 2.014102 | $He^4$ | 4.002603 |
| $H^3$ | <u>3.016050</u> | n | <u>1.008665</u> |
| | 5.030152 | | 5.011268 |

with a mass change of 0.018884 AMU, or 17.59 MeV. The energy calculations for Eqns. 2b and 2d are left as Prob. 1.

Note that the energy released per _reaction_ is much smaller in fusion than in fission, but that the energy released per _nucleon_ is about the same for Eqns. 2a and 2b and is higher for Eqns. 2c and 2d than for fission.

Reaction 2c (the "d-t" reaction) proceeds with greater probabilities and hence at lower energies (temperatures) than reactions 2a or 2b (the "d-d" reaction), for reasons involving the quantum mechanics of nuclear reactions, which we will not go into here. The energy released in the d-t reaction is more than 4 times larger than in the d-d reaction, making a positive energy return easier. These reasons are why the d-t reaction will probably be the one used in the first successful power-producing fusion reactions. However, as we shall see below, the fuel supply for the d-t reaction is limited.

## FUEL SUPPLY

We first consider the supply for the d-t reaction for reasons just given. Tritium ($H^3$) is extremely rare in nature, since it is radioactive with a half-life of only 12.3 years. The only reason it occurs at all is that it is produced by cosmic ray neutrons in the earth's atmosphere by the following reaction

$$N^{14} + n \rightarrow C^{12} + H^3$$

The proposed source of $H^3$ for large scale use of the d-t reaction is from lithium. The simplest reaction is

$$Li^6 + n \rightarrow He^4 + H^3 + 4.78 \text{ MeV} \quad (3a)$$

which also releases energy and proceeds with slow neutrons. An endothermic reaction is

$$Li^7 + n + 2.5 \text{ MeV} \rightarrow H^3 + He^4 + n \quad (3b)$$

The reaction of Eqn. 3b requires fast neutrons to supply the energy, and the final products shown are not the only way a fast neutron can interact. Thus one can not count on $H^3$ production taking place. The neutrons needed for Eqns. 3a and 3b can come from the d-t reaction, Eqn. 2c. Thus a sort of breeding can take place. The breeding ratio can be made greater than one by first having the fast neutrons from the d-t reaction interact with appropriate nuclei in a (n, 2n) reaction. A (n, 2n) reaction means one neutron is incident and two neutrons are emitted.

If only the $Li^6$ reaction is used, the fuel supply is quite limited. Lithium-6 is only 7.42% of naturally occurring lithium. Lithium itself is a relatively rare element, and is readily recovered from only a few deposits and salt brines. While it is always risky estimating mineral resources, the best estimates indicate $1 - 2 \times 10^{13}$ gm of lithium. There are thus perhaps

$$\frac{(.0742)(2\times10^{13}\text{gm})(6.02\times10^{23}\text{atoms/mole})}{6 \text{ gm/mole}} = 1.49\times10^{35}$$

recoverable atoms of $Li^6$. The energy release per $Li^6$ atom is 4.78 MeV (Eqn. 3a) plus 17.59 MeV (Eqn. 2c), which equals 22.37 MeV or $3.58 \times 10^{-12}$ J. The potential energy release from the d-t reaction based on $Li^6$ is thus about $5.3 \times 10^{23}$ J at 100% efficiency. This is only a little less than 3 times the energy content of the world's fossil fuel supply.

If the reaction of Eqn. 3b can be used, then the energy would be increased by at most 8.4 times. This number is less than the ratio of the abundances since Eqn. 3a is exothermic while Eqn. 3b is endothermic. So even with full utilization of all of the lithium, we have only about 21 times the fossil fuel supply for the d-t fusion reaction.

It seems reasonable that if the d-t controlled thermonuclear reaction can be achieved, then the d-d and even the $H^2$ - $He^3$ (Eqn. 2d) reactions can also be done as technology improves. The energy released per deuteron can be obtained by adding the energies of Eqns. 2, and dividing by 6, the number of deuterons on the left. This is because Eqns. 2a and 2b are about equally likely, and the $H^3$ and $He^3$ needed on the left of Eqns. 2c and 2d are end products of the first two reactions. The result is 7.21 MeV/d, or $1.155 \times 10^{-12}$ J/d.

Deuterium is quite plentiful; the oceans are full of it. While $H^2$ is only 0.015% of natural hydrogen, the total is vast. One gram of water has $2(6.02 \times 10^{23})/18 = 6.69 \times 10^{22}$ hydrogen atoms and thus $1.00 \times 10^{19}$ deuterium atoms. The total mass of the oceans is $1.43 \times 10^{24}$ gm. If we suppose that 50% of the $H^2$ can be recovered from the oceans, and that the overall efficiency of the fusion process is 30%, we have

$$(.3)(.5)(1.155x10^{-12}\ J/d)(1.0x10^{19}\ d/gm)(1.43x10^{24}\ gm)$$
$$= 2.5x10^{30}\ J$$

or $1.3x10^7$ times as much as the remaining fossil fuels.

The present world rate of energy usage is about $2x10^{20}$ J/yr, so the supply for the d-d reaction at this rate will last $1.25x10^{10}$ yr, which is probably longer than the earth will be habitable. It is unrealistic to assume the present rate of energy consumption will continue. However, compound growth can not continue either, as the following results indicate. (See Appendix 1 for details.) An annual growth rate of 3% reduces the $1.25x10^{10}$ yr duration down to just 668 years! At the end of that time the power generated would be over 10 000 times the power absorbed from the sun, which is clearly untenable. We will suppose instead that population and power consumption stability set in as in Unit 22 at $16x10^9$ people and $2x10^4$ W per person, which is $3.2x10^{14}$ W or $1.01x10^{22}$ J/yr. The supply would last

$$2.5x10^{30}\ J/1x10^{22}\ J/yr = 2.5x10^8\ yr$$

A quarter billion years is clearly a very long time--long enough to develop some alternative energy source if need be!

## FURTHER DISCUSSION

The temperatures, confinement times, and the techniques to achieve these were covered in Unit 20. We consider here some of the problems of energy recovery and possible adverse environmental effects.

The problems of energy recovery center around the

neutrons produced in Eqns. 2b and 2c. Since they are the lighter reaction product they have a large fraction of the energy. Being neutral they do not interact with electric or magnetic fields thus making direct production of electricity (as by MHD, Unit 19) impossible. We can find the energy of the neutrons as follows. The energies of the initial particles are low enough that we can reasonably assume they are at rest before hand. The final kinetic energies are then equal to the energy released, designated as $Q$:

$$E_{k1} + E_{k2} = Q \tag{4}$$

By momentum conservation, the magnitudes of the final momenta are equal. We can write momentum in terms of kinetic energy as $p = \sqrt{2mE_k}$, so we have

$$\sqrt{2\, m_1\, E_{k1}} = \sqrt{2\, m_2\, E_{k2}} \tag{5}$$

If one solves Eqns. 4 and 5 for $E_{k1}$, the result is

$$E_{k1} = \frac{m_2}{m_1 + m_2}\, Q$$

The neutron thus gets

$$\frac{3}{4}\,(3.27) = 2.45 \text{ MeV}$$

in Eqn. 2b and

$$\frac{4}{5}\,(17.59) = 14.07 \text{ MeV}$$

in Eqn. 2c, where rounded-off numbers were used for the masses. Neutrons thus carry off 16.52 MeV or 38% of the energy of the reactions of Eqns. 2. If the last reaction is excluded, since it involves a doubly charged nucleus and is therefore harder to achieve, then 57% of the energy goes off in neutrons. The neutrons could possibly be used in further nuclear reactions which lead to charged particles. Another way is to let them transfer their energy by elastic collisions. Since the energy transfer is greatest when the mass of the target is the same as the bombarding particle, water is a reasonable substance to use. The water then becomes heated and a conventional steam cycle can be used. The energy of the charged particles will most likely be recovered by MHD.

There is very little adverse environmental impact associated with fusion energy based on the d-d reaction. The "mining" of the fuel involves relatively little energy and leaves no scars as strip mining does. The final products of the reactions are helium isotopes ($He^3$ and $He^4$ if the reaction in Eqn. 2d does not succeed, and $He^4$ if it does). Helium not only is not toxic and causes no change in climate, but is very useful. Thermal pollution should be relatively small because of the very high temperatures of the reactions.

The main problem concerns the possible leakage of radioactive $H^3$ into the biosphere. Since it is chemically identical to ordinary hydrogen, tritium can be incorporated in water and hydrocarbons and thus can find its way into living organisms. If we suppose that there is a one-day inventory of $H^3$ for a 1000 MW (output) power plant using the d-t reaction, and assume 50% overall efficiency, then the activity of that inventory is about $3x10^6$ Ci (see Prob. 3). This activity is much smaller than in the core of a fission reactor which has been operating for a while. The inventory of $H^3$ would be less for a reactor on the d-d cycle. There is the possibility of some accident releasing this much radioactive material. Since it is only being stored, the chances of an accident are much less than for the core of a fission reactor. The main problem is that hydrogen can diffuse through the walls of containers rather easily. The leakage rate can be held down by keeping the container at cryogenic temperatures, and using multiple walls and vacuum pumps. The loss rate could be made as low as 1 Ci/day. For comparison some fission reactions lose over 1000 Ci/day.

The walls of the fusion reactor will become radioactive from neutron capture. Careful choice of materials can lower the capture probability and lead to relatively short half-lives for the induced activity. The problem is no worse than for the walls of a fission reactor, and is many orders of magnitude less than for the radioactive fission fragments.

REFERENCES

See the list at the end of Unit 20.

PROBLEMS

1. Verify the energy released in Eqns. 2b, 2d, and 3a. The masses needed in addition to those already given are $He^3$: 3.016030, and $Li^6$: 6.015125 AMU.

2. Find the energy released in a single laser-pellet micro-explosion. Assume the drop is 0.5 mm in diameter, is composed of deuterium and tritium, has a density of 0.22 $gm/cm^3$, and that 1% of the fuel reacts.

3. Verify that one day's inventory of $H^3$ for a 1000 MW (output) power plant using the d-t reaction is about $3 \times 10^6$ Ci. Assume a 50% overall efficiency.

4. Verify that the full utilization of $Li^7$ results in an increase in the energy supply for the d-t reaction by 8.4.

# Appendix 1
# Compound Growth and Population

## INTRODUCTION

It has been claimed that a large portion of our environmental and energy problems stem from the inability of the public to grasp the implications of compound growth. Here is one example: Suppose that half of the total oil resources have been used up. Since oil has been used for about a hundred years, it is tempting to suppose that it should last for about an equal length of time still. But the fact is that if the oil is being consumed at a rate increasing at 6% per year, all the oil will be gone in just 12 years! Another example of unconstrained growth: roads now cover about 1.5% of the area of the "old 48" of the U.S. If the highways increase by 3% per year, the entire "old 48" will be paved over by the year 2115!

## COMPOUND AND EXPONENTIAL GROWTH

We first consider annual growth rates. Suppose something

increases by p percent per year. We adopt the convention here that p will be quoted as a percentage, but expressed as a decimal in most equations. If the amount to begin with is $A_o$, at the end of one year the amount is

$$A_1 = A_o (1 + p)$$

by the definition of annual growth. The amount at the end of two years is the amount at the end of one year multiplied by $(1 + p)$:

$$A_2 = A_1 (1 + p) = A_o (1 + p)^2$$

We can generalize this to say that the amount after n years is

$$A_n = A_o (1 + p)^n \qquad (1)$$

As an example, consider a savings account which earns 5% per year, with the interest added (or compounded) annually. If one starts with \$100, how much does he have after 20 years?

$$A_{20} = \$100\ (1 + .05)^{20}$$

$$= \$265.33$$

It is clearly tedious to multiply 1.05 by itself twenty times. Some slide rules and a few pocket calculators can give the answer directly. Usually one has to resort to logarithms:

$$\log A_n = \log A_o + n \log (1 + p) \qquad (2)$$

If the question is turned around to ask how long it will take for a given increase, such as a doubling, then the use of logarithms is necessary. Equation 2 may be solved for n to yield

$$n = \frac{\log A_n - \log A_o}{\log (1 + p)}$$

or

$$n = \frac{\log (A_n/A_o)}{\log (1 + p)} \qquad (3)$$

If the number of years for doubling the amount is desired, the above equation becomes

$$n_{double} = \frac{\log 2}{\log ( 1 - p)} = \frac{.30103}{\log (1 + p)} \qquad (4)$$

In our previous example, the time needed for the money to grow to \$200 is

$$n_{double} = \frac{.30103}{\log (1.05)} \approx 14.2 \text{ yr.}$$

The doubling times for a variety of annual percentage growth rates are given in Table 1. An interesting point emerges from a close examination of Table 1. If the percentage and the doubling times are multiplied together, the result is always near 72 (ranging from 69.6 to 76.0, if the 30% figure is omitted). Thus one has as a close approximation (for $p$ as a percent):

$$n_{double} = \frac{72}{p} \qquad (5)$$

Table 1. Doubling times for various growth rates.

| p | 1% | 2 | 3 | 4 | 5 | 6 | 7 | 8 |
|---|---|---|---|---|---|---|---|---|
| $n_{double}$ | 69.6 yr | 35.0 | 23.4 | 17.7 | 14.2 | 11.9 | 10.2 | 9.0 |

| p | 9% | 10 | 12 | 15 | 20 | 30 |
|---|---|---|---|---|---|---|
| $n_{double}$ | 8.0 yr | 7.3 | 6.1 | 5.0 | 3.8 | 2.6 |

Equation 5 is the most useful equation in any question of growth known to the author. The basis of Eqn. 5 is rather easy to explain. We return to Eqn. 4, but use natural logarithms instead:

$$n_{double} = \frac{\ln 2}{\ln (1 + p)} = \frac{.69315}{\ln (1 + p)}$$

There is a well-known approximation that $\ln (1 + x) \simeq x$ for small x. If this is used, and p converted to a percent, one obtains

$$n_{double} = \frac{69.3}{p}$$

The approximation $\ln (1 + x) \simeq x$ is the first term in a series expansion. For larger values of x, other terms become important. Equation 5 has 72 in the numerator instead of 69.3 to partially compensate for this.

As an example, world population is currently growing at about 2% per year. How long will it take to double?

$$n_{double} = 72/2 = 36 \text{ yr}$$

The concept can be extended to periods other than doubling by finding the closest power of 2. For example, suppose that the largest supportable population of the world is assumed to be $30 \times 10^9$ people. How long will it take to reach that at the present growth rate? The increase is about 7.8 times the present population, or about $2^3$. It will thus take a bit less than 3 doubling periods, each 36 yr long, or only about 108 years. Usually this method yields results with sufficient accuracy, since there is considerable uncertainty in making future projections. There is a Table of powers-of-two included as Appendix 5 for convenience. A good approximation is that

$$10^n \simeq 2^{3.32\,n}$$

Growth of something like a population is continous. It does not occur in one sudden jump once every year. Even most savings accounts accumulate more often than annually: every 3 months, monthly, or even daily. There are two ways to characterize growth rates on a more continuous basis. One is the annualized rate, which is the actual growth during the shorter period multiplied by the number of times that period goes into a year. For example a monthly interest of 1.5% means an annualized rate of (1.5)(12) = 18% per year. We suppose that the year is divided into $m$ periods with an annualized rate of $p'$. The growth during one of the periods is

$$(1 + p'/m)$$

and the growth during one year (which is $m$ periods) is

$$\left(1 + \frac{p'}{m}\right)^m$$

To illustrate the difference between an annualized growth rate and the actual annual increase, consider the following example. Suppose something is growing 2% per month. The annualized rate is thus 24%, but the actual growth in one year is

$$(1.02)^{12} = 1.2682$$

or about 27%.

For something like a population, growth is continuous and one should divide the year into an unlimited number of periods. In other words we seek

$$\lim_{m \to \infty} \left(1 + \frac{p'}{m}\right)^m$$

This is a well-defined and well-known limit:

$$\lim_{m \to \infty} \left(1 + \frac{p'}{m}\right)^m = e^{p'}$$

Thus continuous growth at an annualized rate of $p'$ is exponential growth with a yearly increase of $e^{p'}$. If $n$ years of growth occur, the result is $(e^{p'})^n = e^{np'}$.

The other way of referring to continuous growth is in terms of the actual annual increase in one year, which is the annual rate $p$. The two rates $p$, and $p'$ are not the same, as the example given above shows. We can relate the annual increase $p$ and the annualized rate $p'$ for continuous growth as follows. A series expansion of $e^{p'}$ is:

$$e^{p'} = 1 + p' + (p')^2/2 + (p')^3/6 + \cdots$$

The sum of the series is the actual final amount after one year. The final amount with an annual increase is just $(1 + p)$. If $p'$ is known, $p$ is found by

$$p = e^{p'} - 1$$

If $p$ is known, $p'$ may be found by

$$p' = \ln(1 + p)$$

If $p'$ is small enough that the quadratic and higher terms in the series expansion can be neglected, the $p$ and $p'$ are approximately the same. As a rough rule, the annual and annualized rates are close enough the same up to about 10%.

An alternative way to consider continuous growth is by setting up a differential equation. Let $dA$ be the increase in the amount during time $dt$. In any growth situation the increase is proportional to the amount already present, $A$, and the length of time, $dt$. Thus we have

$$dA \propto A\,dt$$

We let the constant of proportionality be $\lambda$, which thus represents the growth per unit time, to obtain

$$dA = A\,\lambda\,dt$$

This equation may be solved by separation of variables to yield

$$A = A_o\,e^{\lambda t}$$

where $A_o$ is the amount when $t = 0$. Note that by letting $\lambda$ be negative, one can accomodate continuous decay, as in Unit 18.

SOME CONSEQUENCES OF GROWTH

We first prove that with continuous growth the amount of a product consumed during one doubling period is equal to the total used for all time up to the beginning of the doubling period in question. The total amount used during a time interval $t_1 \rightarrow t_2$ is the area under the rate of usage curve. The rage of usage increases exponentially for continuous growth. The total amount used up to $t_1$ is thus

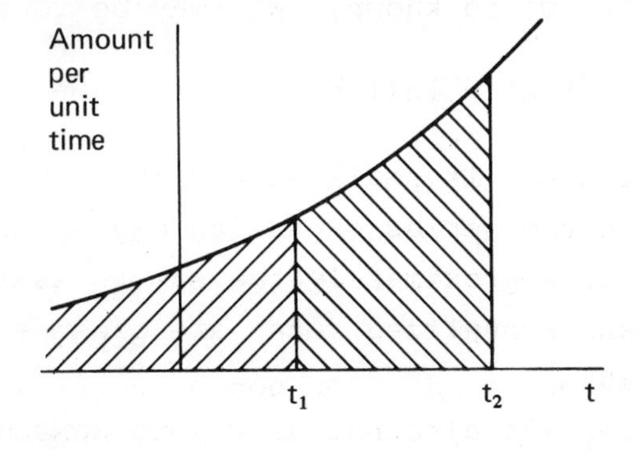


$$\int_{-\infty}^{t_1} A_o\,e^{\lambda t}\,dt = A_o\,\frac{e^{\lambda t}}{\lambda}\Bigg|_{-\infty}^{t_1} = \frac{A_o}{\lambda}\,e^{\lambda t_1}$$

The total amount used during $t_1$ to $t_2$ is

$$\int_{t_1}^{t_2} A_o e^{\lambda t} dt = \frac{A_o}{\lambda}(e^{\lambda t_2} - e^{\lambda t_1})$$
$$= \frac{A_o}{\lambda} e^{\lambda t_1} (e^{\lambda(t_2 - t_1)} - 1)$$

If the interval $t_1 \rightarrow t_2$ is a doubling period, we have

$$e^{\lambda t_2} = 2 e^{\lambda t_1}$$

or

$$e^{\lambda(t_2 - t_1)} = 2$$

Thus the amount used during the doubling period $t_1 \rightarrow t_2$ is

$$\frac{A_o}{\lambda} e^{\lambda t_1} (2 - 1) = \frac{A_o}{\lambda} e^{\lambda t_1}$$

which is the same as the total used during all time up to $t_1$. The following may help clarify this result, since it is sometimes hard to believe the result. Consider the series

1, 2, 4, 8, 16, 32, .....

where every term is double the one before it. Any given term is always one more than the sum of all of the prior members of the series.

Thus when one has some resource, such as oil, being consumed at approximately an exponentially growing rate, and one has arrived at the point in time where half of that resource has been consumed, it is time (or past time!) for alarm, since there is only one doubling period left before it is all gone. This result is the basis for the claim in the opening paragraph about oil. A 6% increase per year implies a doubling time of 12 years.

A false security often exists about the question of

population growth. The incorrect feeling goes as follows. Suppose it has been determined that only 30 billion people can live on earth with some reasonable standard of living, and that the population of the world at some future time is 15 billion. The error is to think that since it took maybe 100 000 years to achieve the first 15 billion, then there will be a long period needed to achieve the next 15 billion and thus there is plenty of time before one need worry about limiting population. In actual fact one more doubling period, which is around 36 years at present rates, will do it! Planning for stability will take longer than that.

It is useful to have an expression for how long a resource will last if its rate of consumption shows compound growth. Although growth of consumption is more or less a continuous process, we will use annual compound growth, since the statistics are usually given on an annual basis. During the first year an amount $A_o$ will be used. The second year it will be $A_1 = A_o (1 + p)$, the third year $A_2 = A_o (1 + p)^2$, and so on. The total amount used in N years is

$$A_o + A_o (1 + p) + A_o (1 + p)^2 + .... + A_o (1 + p)^{N-1}$$

$$= \sum_{n=o}^{N-1} A_o (1 + p)^n$$

This sum must be equal to the total supply, S:

$$S = \sum_{n=o}^{N-1} A_o (1 + p)^n \qquad (6)$$

We must solve Eqn. 6 for N. This task is not too difficult, since we are summing a geometrical series, and the summation can be written in closed form as

$$S = A_o [(1 + p)^N - 1]/p$$

This equation can be rearranged to yield

$$(1 + p)^N = \frac{Sp}{A_o} + 1$$

We take the logarithm of both sides to obtain

$$N = \log \left( \frac{Sp}{A_o} + 1 \right) \Big/ \log (1 + p) \qquad (7)$$

where $N$ is the number of years a given resource will last if the present supply is $S$, the current rate of use is $A_o$, and the rate of use is growing at $p$ percent per year. In Eqn. 7, $p$ is to be expressed as a decimal. The ratio $S/A_o$ is the number of years the resource would last at the present rate of use, $A_o$. It can be shown that Eqn. 7 reduces to this result in the limit $p \to 0$. A result similar to Eqn. 7 can be derived by integrating continuous growth. See Problem 4.

A final remark on compound growth: it simply can not go on forever. Some natural limitation sets in sooner or later. In the case of consuming a resource, as the resource becomes scarcer, the supply decreases. The price will rise and the demand will fall, giving rise to a curve

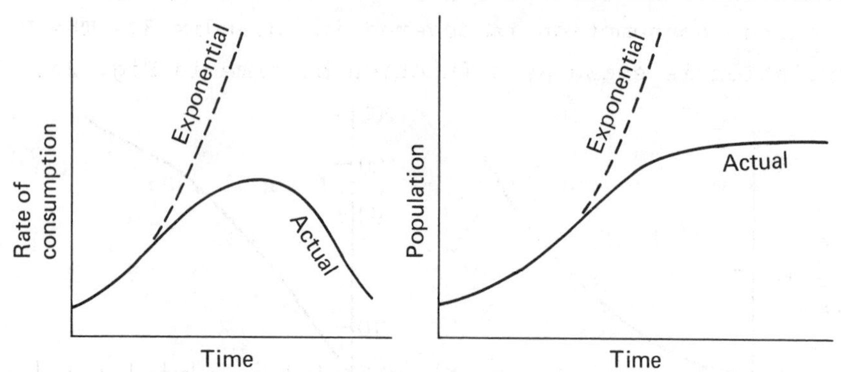


Fig. 1. (a) the consumption of a depletable resource as a function of time. (b) the growth of a population.

like Fig. 1a. In the case of populations, there is some limit such as food supply which results in the population leveling off, as in Fig. 1b, or even decreasing.

Yet many people still believe in compound growth without limit. One of the best ways to show that somewhere population, energy consumption, economy, cars, etc., must level off is to take the compound growth to some obviously ridiculous level. One example was already given, in the case of roads completely covering the U.S. Other examples are a layer of people 100 ft thick covering the entire globe, and power consumption equal to that produced by the sun. Everyone will agree that these limits are indeed beyond what is practical. But the surprising aspect is that these ridiculous limits are reached in relatively short times if compound growth continues. See some of the problems. The average person simply does not understand how quickly compound growth builds up. In some cases the limit is reaches soon enough, and the time needed to achieve a change is long enough, that we must start now.

## POPULATION GROWTH

We consider the growth of population as an example. Growth of energy consumption is covered in Appendix 3. The U.S. population is shown as a function of time in Fig. 2a. It

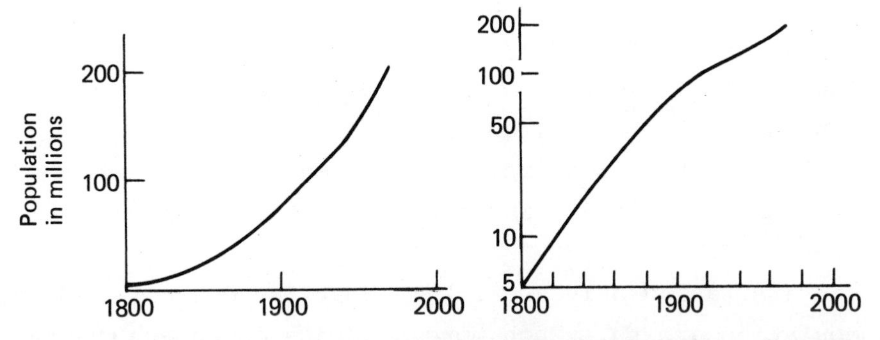


Fig. 2. (a) linear and (b) semi-log graph of U.S. population as a function of time.

is clear that the population is growing, but is it compound growth? In the limit of continuous growth, we have an exponential dependency:

$$N = N_o e^{p't}$$

If we take the natural logarthm of both sides we have

$$\ln N = \ln N_o + p't$$

Thus the logarithm of a quantity having compound growth is a linear function of time. Such a graph is called a semi-log graph, and is shown for the U.S. population in Fig. 2b. We see that the data do not fall on a single straight line, indicating that the population has not been following compound growth at a single rate. However, approximately one straight line fits from 1800 to 1860 and another from 1910 to 1970, with a smaller slope and hence a slower rate of growth. A semi-log graph is the best way to test for compound growth. If the points lie reasonably well along a single straight line, then the growth is compound at a constant rate. There usually will be some minor deviations which must be smoothed over by a "best fit" straight line. The slope of this line gives the growth rate. The best way to find the growth rate from the line is to find the doubling time and then use Eqn. 5.

Some figures on world population are given in Table 2. The important thing to note is that the doubling times have been decreasing, indicating an increasing rate of growth. The last doubling time given in Table 2 (estimate to 4 billion) will take 45 years. The present rate of increase is 2% per year, indicating a doubling time of only 36 years.

We are taking only a quick look at the overall picture of population growth. Much more accurate demographic projections can be made. The age distribution of the present

Table 2. World Population at Different Times

| Year | Population |
|---|---|
| 1650 | 0.5 billion |
| 1800 | 1 billion |
| 1930 | 2 billion |
| 1975 (est) | 4 billion |

population, the male/female ratio, trends in birth rate and mortality as a function of age, and so on, must be considered. See Refs. 1 and 2 for some more details.

## LIMITS TO POPULATION

The basic factors which limit the population are food availability, living space, energy and other resources, and pollution levels.

There are about $8 - 16 \times 10^9$ acres available and suitable for food production. One can consider the rate at which energy from the sun reaches the land, and the rather low efficiency for the energy conversion by photosynthesis and the food chain to arrive at the result that about 1 A is needed to feed one person. These figures imply a world population of $8 - 16 \times 10^9$ people. If one allows for some improvement in agriculture, for the use of more fertilizer, and more irrigation, this figure might be raised to $30 \times 10^9$ people. It must be noted that energy and the supply of raw materials place a definite limit to additional irrigation and fertilization. The present (1973) world population is about $3.85 \times 10^9$ people. Thus the above limit of $30 \times 10^9$ people is less than 8 times, or 3 doubling periods, more than present. If the current rate of growth of 2% per year holds, three doubling periods will take only 108 years. Thus the ultimate population might be reached by 2081.

Suppose for the moment that technology succeeds in increasing the above limit by a factor of 8 (vegetarian diets help increase the limit). In that case the time for attaining the ultimate population is put off only another 108 years, or until 2189. At that time the total population would be $240 \times 10^9$ people. Since the land area of the world (including all artic, jungle, and mountainous regions) is about $5.8 \times 10^7$ $mi^2$ or about $3.7 \times 10^{10}$ acres, there would be about 6.5 people per acre.

The limits to energy consumption are based on resources and on climate alterations. Since most energy ultimately ends up as heat, too much energy used will warm up the climate. When energy use amounts to about 7% of the solar energy absorbed by the earth, the mean temperature will rise about 5 °C. See Unit 23 and Appendix 3 for further discussions on energy.

Resources may not place much of a constraint on population _if_ there is enough energy for recycling them. With only very minor exceptions, no material is lost from the earth, and in principle all could be reused. If the inventory in use is large enough and the supply small enough, there may be a limitation to expanded usage.

Right now it is difficult to say what the effects of pollution may be on population, since it depends critically on how well present sources can be controlled and what new low-pollution processes might be developed.

REFERENCES

1. A. Turk et al., _Ecology, Pollution, Environment_ (W. B. Saunders, 1972) Ch. 8.

2. J. Harte and R. H. Socolow, _Patient Earth_ (Holt, Rinehart, and Winston, 1971) Appendices 1 and 2.

3. D. H. Meadows et al., _The Limits to Growth_ (Potomac Associates, 1972).

PROBLEMS

1. Assume the present population of the world and growth rates as given above. How long would it take to have
   a) 1 person per square foot of land area
   b) a layer of people 100 ft thick over the entire globe?

2. U. S. power consumption is now about $2.2\times10^{12}$ W, and growing about 3% per year. Assume that the average solar power absorbed by the earth (including the atmosphere) is 900 W/m$^2$ during the sunlit hours. How long will it be before the U.S. power consumption reaches 7% of the solar energy absorbed by the U.S.? The land area is $3.6\times10^6$ mi$^2$.

3. The world rate of energy consumption is about $6.8\times10^{12}$ W and growing about 3% per year. How long before it equals the sun's total output of $3.9\times10^{26}$ W?

4. In deriving Eqn. 7 we assumed annual compound growth. An alternative approach is to assume continuous (exponential) growth. By integrating the rate-of-usage expression, derive

$$N = \frac{1}{p'} \ln \left(\frac{p's}{A_o} + 1\right)$$

   and show that for p' small Eqn. 7 reduces to this result.

5. Make a semi-log graph of the world's population. What can you conclude?

6. One can use the ideas of growth to extrapolate backwards in time. Assume the present world population is $3.8\times10^9$ people and that it has always been

growing at 2% per year (not actually true). Extrapolate back in time to the point when there would have been only 2 people under the assumption of constant growth. The answer again points out that population is growing gaster than at a compound rate.

# Appendix 2
# Estimating

## INTRODUCTION

Frequently exact data or answers are not available in environmental questions. In other cases an approximate result will suffice. For example, a rough calculation will often indicate whether a particular effect is or is not significant. In the above cases an ability to estimate is needed. The skill of estimating is seldom included in formal classwork. There are no complete rules, but some guidelines and suggestions can be given. As in many things, practice is the best way to learn.

The best starting point in estimating is to relate the unknown to something that is known. If you do not know anything, make an intelligent guess. Try to bracket your guess. That means have a number you are rather sure the quantity in question is greater than and another you feel it is less than. Sometimes even being off by two orders of magnitude will be sufficiently close. Always keep in mind that in estimating you are dealing with approximate numbers, so do not carry out arithmetic operations to very many significant figures. Usually the nearest integer is good enough. Occasionally even a round-off to the nearest power-of-ten will suffice. In rounding to the nearest power-of-ten, think of it logarithmically. That is, 1 to 3 rounds down, and 4 - 9 rounds up. Some short-cut arithmetical approximations help. One of the most useful is

$$(1 + x)^n \simeq (1 + n x)$$

for $x$ less than about .2 or .3. In particular,

$$(1 \pm x)^2 \simeq 1 \pm 2x$$

$$\frac{1}{1 \pm x} \simeq 1 \mp x$$

$$\sqrt{1 \pm x} \simeq 1 \pm x/2$$

Some other approximations are

$$2^n \simeq 10^{.3n}$$

$$e^x \simeq 10^{.4x}$$

$$\ln (1 + x) \simeq x \quad \text{for } x \lesssim .2$$

$$4\pi \simeq 10$$

$$N \pm n \simeq N \quad \text{for } N >> n$$

$$\text{Number of seconds in a year} \simeq \pi \times 10^7$$

For estimates to work satisfactorily, you must avoid situations in which you are seeking the difference of two nearly equal estimated numbers. Frequently the problem can be rearranged to avoid this.

Since experience and practice are the best teachers, several examples are now given. The first example is to find the mass of the atmosphere in kilograms. A known starting point is that the atmospheric pressure is 14.7 $lb/in.^2$ at sea level. This is the weight per square inch. Thus if we find the number of square inches of surface area of the earth we have the total weight, which can be converted to mass by 1 lb $\sim$ .5 kg. Most people remember the diameter of the earth is about 8000 miles. So we have:

$$
\begin{aligned}
\text{Weight} &= (\text{pressure})(\text{area in square inches}) \\
&\simeq (15)(4\pi)(4000)^2(5280)^2(12)^2 \\
&\simeq (15)(10)(15000000)(25000000)(150) \\
&= (1.5)(1.5)(2.5)(1.5)\text{x}10^{18} \\
&\simeq 8\text{x}10^{18} \text{ lb} \\
&\sim 4\text{x}10^{18} \text{ kg}
\end{aligned}
$$

The actual answer is $5.1\text{x}10^{18}$ kg, so we are adequately close for many purposes. (In this particular case one could have used better input data and have done exact arithmetic to obtain an answer which is very close.)

A similar calculation, but one involving some guessing, is finding the mass of the oceans in kilograms. We know the density of sea water is about 1 $gm/cm^3$. So we must find the volume of the oceans. Oceans cover about 2/3 of the surface. We can bracket the depth by knowing that most of the ocean is more than 1000 ft deep and less than 5 miles (25000 ft) deep. We pick 10000 ft. The volume is thus

$$
\begin{aligned}
V &= (\text{fraction of area})(\text{total survace area})(\text{depth}) \\
&\simeq (.6)(4\pi)(4000)^2\ (5280)^2(10000)
\end{aligned}
$$

$$\simeq 3\text{x}10^{19} \text{ ft}^3$$

Now there are about 30 cm to a foot, or about $(30)^3 \simeq 3\text{x}10^4$ $\text{cm}^3$ to a cubic foot. Thus we have a volume of $10^{24}$ $\text{cm}^3$, or a mass of $10^{24}$ gm or $10^{21}$ kg. The actual answer is $1.43\text{x}10^{21}$ kg, so we again have a reasonable estimate.

The above estimates would be useful for finding the approximate pollution concentration if a given amount of pollutant is released in the air or oceans, for example.

Here is an example in line with the way things arise in environmental questions. How complete must the cumbustion in the engine of a car be to meet the 1975 standard of 3.5 gm of CO per mile? We obtain the answer by first finding how much $CO_2$ is produced assuming complete combustion. We may start with the knowledge that a typical car obtains 8 to 25 mi/gal. We will take 12 mi/gal. This means that 1/12 gal is consumed per mile. Next we look up some data on gasoline in the Handbook of Chemistry and Physics. We find that gasoline is about 85% carbon (by weight), and one gallon weighs 6 lb. Thus in one mile 6(1/12) or 0.5 lb of gasoline is consumed. The carbon in that weighs 0.4 lb which implies a mass of 0.2 kg. If we suppose complete combustion, the final product is $CO_2$. Carbon is 27% of $CO_2$. So $.2/.3 \simeq .7$ kg = 700 gm of $CO_2$ is produced per mile. Thus 3.5 gm/mi of CO implies that the combustion must be 99.5% complete.

The foregoing examples are meant to illustrate how one can start with known data, perhaps estimating unknown information or looking up some additional figures, and arrive at reasonable estimates. A few problems with starting hints are given below for additional practice.

PROBLEMS

1. How much habitable land is there in the world? Hints: start from the radius of the earth to obtain the total surface area of the earth. Use the approximate number that land is about 1/3 of the surface area, and estimate what fraction of the land is habitable--take a look at a globe.

2. How much air is breathed by a person in one day? Hints: what is the volume per breath? Either measure this by breathing through a straw into a cup inverted in water or guess it from experience blowing up a baloon. Then estimate the total number of breaths in a day.

3. What fraction of land area in the U.S. (excluding Alaska) is devoted to roads? Use the fact that there are about 4 million miles of roads. Hints: estimate a reasonable average width for the total road and adjacent right-of-way. Estimate the total land area of the U. S. from approximate dimensions.

4. What is the average daily run-off of rain in the U. S., excluding Alaska? Hints: what is the total land area? What is a good figure for the average annual rainfall in inches? Assume some reasonable fraction of the rain soaks into the land and does not run off.

# Appendix 3
# Energy Consumption and Resources

## INTRODUCTION

The use of energy sources other than animal muscle is extremely important in modern society. There is a strong (but not perfect) correlation between the standard of living as measured by the per capita gross national product and the per capita energy consumption. See the article by Cook in Reference 2. Complete recycling of materials is possible if sufficient energy is available. On the other side of the coin is the fact that most of our environmental problems can be related directly or indirectly to the use of energy. The fossil fuel supply is clearly finite and can be exhausted. Even if other energy sources are developed in large scale, there is an ultimate limit to the rate of energy usage based on considerations of changes in climate.

In this Appendix we look at the growth of total energy consumption, the changing patterns in the sources of the energy, and the size of the supply of several important energy sources. Full discussion of these points would be much too lengthy, so mostly a summary of important points is presented. In particular the problems and ways of estimating fossil fuel supplies are not covered here. See Reference 1.

We mostly focus on the statistics for the U.S.A. for several reasons. The U.S.A. has the highest per capita consumption of energy and therefore is experiencing some of the greatest problems. Some of the world energy data is hard to obtain, while it is easily available for the U.S.A. Finally this book will probably be used mostly in the U.S.A.

## U. S. ENERGY CONSUMPTION

The total annual energy and per capita annual energy figures are given in Table 1 and drawn in Fig. 1. Since energy per year is a statement of power, we express the results in terms of Watts. Note that Fig. 1 is a semi-logarithmic graph to illustrate that the growth of total power usage and of per capita power usage in recent years has been approximately at a steady rate (see Appendix 1). Similar figures are given in Table 2 and Fig. 2 for the electricity produced for public use. The average rate of growth of total annual energy usage over the period 1850 to 1970 is 2.8%/yr. The growth rate for total electricity from 1920 to 1970 is about 7.7%/yr. The shifting in the sources of energy is shown in Fig. 3.

Table 1. U.S. Energy per Year

| Year | Total energy/year | Energy/year/person |
|---|---|---|
| 1850 | $7.5 \times 10^{10}$ W | 3250 W |
| 1900 | 25 | 3300 |
| 1920 | 60 | 5700 |
| 1940 | 75 | 5725 |
| 1950 | 110 | 7300 |
| 1960 | 150 | 8400 |
| 1970 | 210 | 10300 |

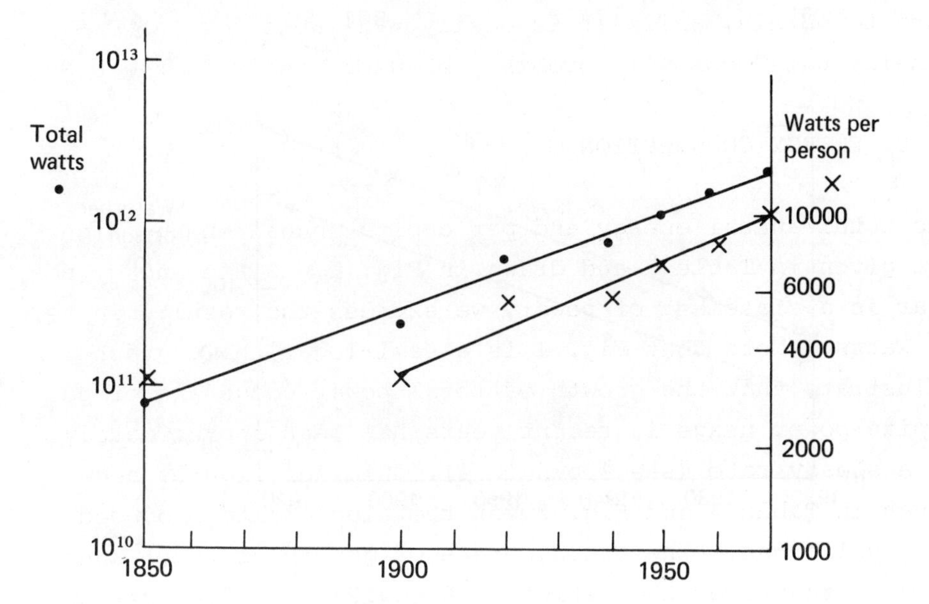


Fig. 1. Total (• left scale) and per capita (× right scale) consumption of electricity in the U.S.

Table 2. U.S. Electricity Production

| Year | Total | Per Person |
|---|---|---|
| 1920 | $4.6\times10^9$ W | 44 W |
| 1930 | 10.2 | 83 |
| 1940 | 14.8 | 113 |
| 1950 | 37.5 | 248 |
| 1955 | 62.4 | 378 |
| 1960 | 85.9 | 480 |
| 1965 | 120.4 | 622 |
| 1970 | 174.8 | 861 |

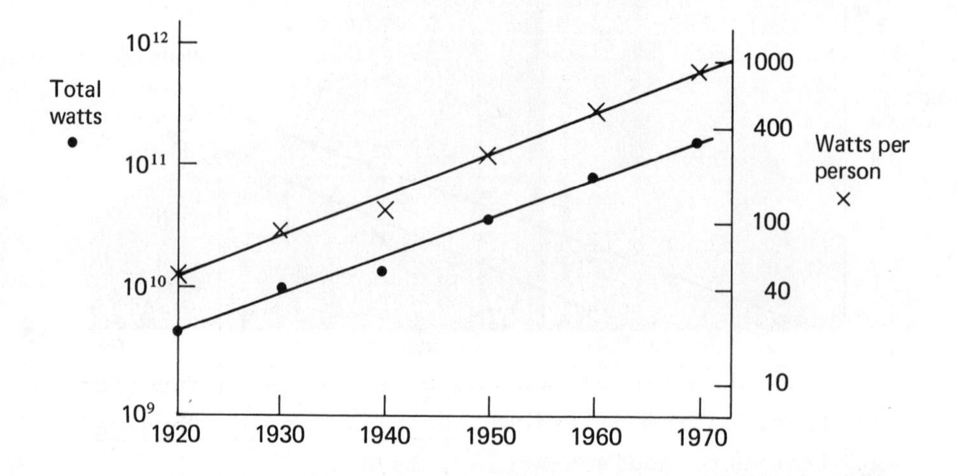


Fig. 2. Total (• ) and per capita (× ) consumption of electricity.

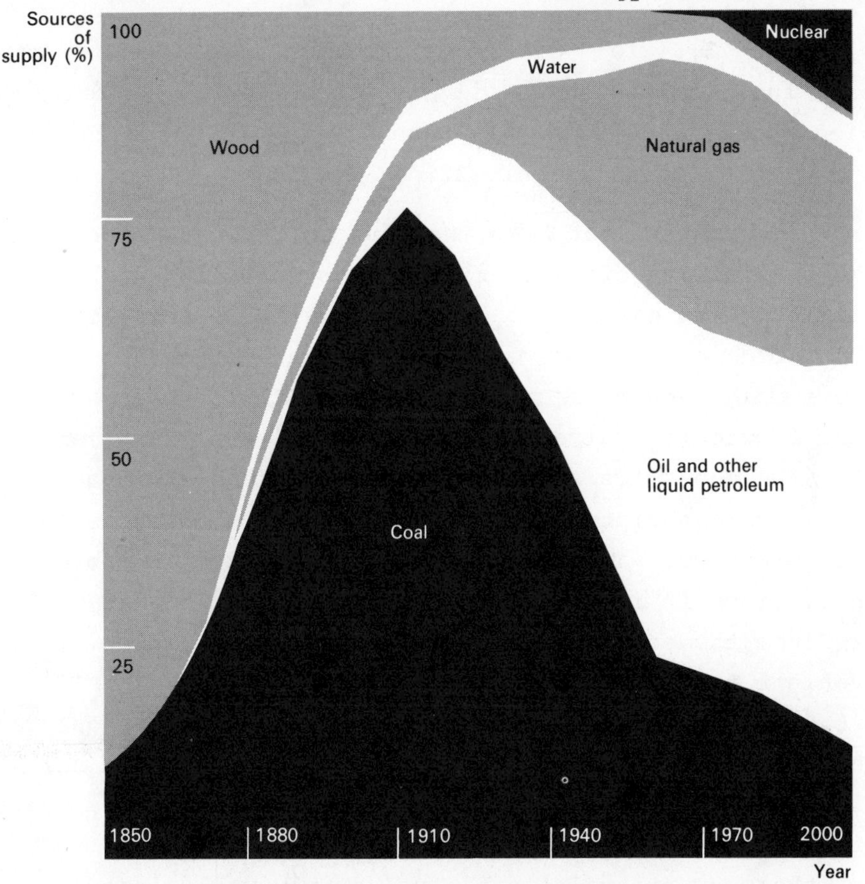


Fig. 3. Fractional sources of energy as a function of time. (From J. Priest, Problems of Our Physical Environment, Addison-Wesley, 1973).

## U.S. ENERGY SUPPLY

It is always hazardous to estimate amounts of buried resources. Experts do not agree with each other. We give here some of the middle-ground figures for fossil fuels. See Table 3. The amount remaining is given in the common units used for that fuel, and also converted to Joules of energy content for ease of comparison. The current rate of usage of the fuel is also given. It should be noted that some estimates show about twice the coal, ten times the oil, and four times the natural gas as given in Table 3. The energy total of coal, oil, and gas in

Table 3 is $3.64 \times 10^{22}$ J. It is clear that coal is our major fossil fuel (unless shale oil extraction becomes a reality). Nuclear power and its supply were covered in Units 26 and 28, and solar power in Unit 22.

Shale oil has not been included in Table 3, since the total feasibility of recovering it has not been established. Preliminary estimates of the amount available show perhaps $8 \times 10^{10}$ bbl readily recoverable in North America and a possible supply ranging from $2 \times 10^{12}$ to $5 \times 10^{13}$ bbl. The higher figure is equivalent to $3 \times 10^{23}$ J, well surpassing coal. The economics and most importantly the environmental destruction associated with recovering shale oil have yet to be worked out. Tidal power is not included in Table 3 either, since it is quite minor as was found in Unit 9.

Since the energy content is expressed in Joules, it is convenient to convert the U.S. total power from $2.1 \times 10^{12}$ W to $6.6 \times 10^{19}$ J/yr. Thus at the present rate of use our estimate of the U.S. fossil fuel supply (excluding shale oil, nuclear, geothermal, and solar power) will last 551 years. If the growth in the use of power continues at 2.8% per year, the conventional fossil fuel supply would last

$$N = \frac{\log [551 (.028) + 1]}{\log (1.028)} = 101 \text{ yr}$$

where Eqn. 7 of Appendix 1 was used.

Table 3. Supply of Certain U.S. Energy Sources

| | Estimated Amount | Energy Content | Current Annual Production |
|---|---|---|---|
| Coal | $1.2 \times 10^{12}$ tons | $3.3 \times 10^{22}$ J | $6.1 \times 10^{8}$ tons |
| Oil and natural gas liquids | $2.7 \times 10^{11}$ barrels | $1.7 \times 10^{21}$ J | $4.1 \times 10^{9}$ bbl |
| Natural gas | $1.6 \times 10^{15}$ ft$^3$ | $1.7 \times 10^{21}$ J | $2.2 \times 10^{13}$ ft$^3$ |
| | Present Installed | | Potential |
| Hydropower | $4.5 \times 10^{10}$ W | | $1.6 \times 10^{11}$ W |
| Geothermal power | $2 \times 10^{8}$ W | | $10^{10}$ to $10^{12}$ W |

The estimated amounts in Table 3 may prove to be too conservative, or in some cases a bit too optimistic. The important point is that the fossil fuel resources listed are definitely finite and will be exhausted soon. Even if the supply of fossil fuels is underestimated by a factor of 8 in Table 3, there is not much time left if compound growth continues. A 2.8%/yr growth rate implies a doubling time of 25 years. So a factor of 8 increase will add only another 75 years to the duration of the supply. (See Appendix 1.) It takes around 50 years for a substantial switch in the method of producing energy. So while we are not going to run out immediately, we must start developing energy alternatives now.

Part of the present energy problem is connected with air pollution control. Coal is definitely the dirtiest fuel. In an effort to reduce air pollution many users switched from the most plentiful fossil fuel to gas and oil, helping to create shortages in the supply of these fuels. Another thrust to clean the air was to equip automobiles with control devices which generally increased fuel consumption. An additional environmental consideration is directed to reducing the effects of strip mining of coal. In many cases the result was stoppage of mining rather than reclamation. Thus the earlier efforts to improve the environment were undertaken without complete consideration of the energy impact. The total picture must now be considered: adequate energy <u>and</u> a good environment. These twin goals can be achieved with present and developing technologies, but more research and development are needed.

## WORLD ENERGY PICTURE

The use of power is very unevenly distributed about the globe. The U.S.A. has the highest per capita use, about $10^4$ W per person. With only 6% of the world's population, we use about a third of the world's energy. Current total world power usage is about $6.4 \times 10^{12}$ W or $2 \times 10^{20}$ J/yr. Some of the world energy resources are given in Table 4. There is considerably greater uncertainty in the Table 4 figures than for those in Table 3, since the U.S. has been much more extensively explored for resources than many other parts of the world. Shale oil estimates for the world show perhaps $1.9 \times 10^{11}$ bbl readily recoverable and a possibility of $1 \times 10^{13}$ to $3 \times 10^{14}$ bbl. The total energy content of the fossil fuels given in Table 4 is about $2.2 \times 10^{23}$ J. At the present rates of using energy, this supply will last 1100 years. If the usage grows at 3% per year, this supply will last only 119 years.

Table 4. Supply of Certain World Energy Sources

| | Estimated Amount | Energy Content | Current Annual Production |
|---|---|---|---|
| Coal | $7 \times 10^{12}$ tons | $1.9 \times 10^{23}$ J | $3 \times 10^{9}$ tons |
| Oil | $2.5 \times 10^{12}$ bbl | $1.6 \times 10^{22}$ J | $1.7 \times 10^{10}$ bbl |
| Natural gas | $1 \times 10^{16}$ ft$^3$ | $1.1 \times 10^{22}$ J | $8 \times 10^{13}$ ft$^3$ |

| | Present Installed | Potential |
|---|---|---|
| Hydropower | $1.5 \times 10^{11}$ W | $2.8 \times 10^{12}$ W |
| Geothermal power | $1.2 \times 10^{9}$ W | $10^{11}$ to $10^{13}$ W |

REFERENCES

1. M. K. Hubbert, in Resources and Man (W. H. Freeman, 1969) Ch. 8.

2. The entire Sept. 1971 issue of Scientific American. Also available as Energy and Power (W. H. Freeman, 1972).

3. A. L. Hammond et al., Energy and the Future (American Association for the Advancement of Science, Washington D. C., 1973).

4. J. C. Fisher, Energy Crises in Perspective (Wiley, 1973). A summary appeared in Physics Today 26 No. 12, 40 (Dec. 1973).

PROBLEM

If the rate of energy consumption decreases, the supply of fossil fuels will last longer. Find what annual rate of decrease will make the world's fossil fuel supply given in Table 4 last forever. See Unit 18 and Appendix 1. The answer is surprisingly small.

# Appendix 4
# Conversion Factors and Other Useful Information

Table 1. Length, Area, Volume, Mass, and Weight

1m = 39.37 in. = 3.281 ft = $6.214 \times 10^{-4}$ mi
1km = .6214 mi = 3281 ft
1 in. = 2.54 cm
1 ft = .3048 m
1 mi = 1609 m = 5280 ft
1 $m^2$ = 10.76 $ft^2$ = 1550 $in.^2$ = $2.47 \times 10^{-4}$ A
1 $ft^2$ = $9.29 \times 10^{-2}$ $m^2$
1 $in.^2$ = $6.452 \times 10^{-4}$ $m^2$
1 $mi^2$ = $2.788 \times 10^{8}$ $ft^2$ = 640 A = $2.59 \times 10^{6}$ $m^2$
1A = 43560 $ft^2$ = 4047 $m^2$
1 hectare = $10^4$ $m^2$ = 2.471 A
1 $m^3$ = 35.31 $ft^3$ = 264.2 gal (US)
1 $ft^3$ = 7.481 gal (US) = .0283 $m^3$
1 gal (US) = $3.78 \times 10^{-3}$ $m^3$ = 3.78 liter
1 $mi^3$ = $1.472 \times 10^{11}$ $ft^3$ = $1.10 \times 10^{12}$ gal (US)
= $4.17 \times 10^{9}$ $m^3$

1 bbl (oil) = 42 gal (US) = .159 $m^3$

1 A-ft = 43560 $ft^3$ = $3.259 \times 10^5$ gal (US) = 1233 $m^3$

1 kg ↔ 2.205 lb (mass-weight at standard gravity)

1 lb ↔ .4536 kg

1 short ton = 2000 lb ↔ 907.2 kg

1 metric ton = 1000 kg ↔ 2205 lb

1 AMU = $1.660 \times 10^{-27}$ kg

Table 2. Rates, Force, and Pressure

1 mph = 1.4667 ft/sec = .447 m/sec = 1.609 km/hr

1 m/sec = 3.281 ft/sec = 3.6 km/hr = 2.237 mph

1 km/hr = .2778 m/sec

1 ft/sec = .3048 m/sec = .6818 mph

1 $ft^3$/sec = 448.8 gal/min = .0283 $m^3$/sec

1 N = .225 lb

1 $N/m^2$ = $1.45 \times 10^{-4}$ lb/$in.^2$ = $7.50 \times 10^{-4}$ cm-Hg

1 lb/$in.^2$ = 6895 $N/m^2$

1 atmos = 14.7 lb/$in.^2$ = $1.013 \times 10^5$ $N/m^2$

1 lb/$ft^2$ = $6.94 \times 10^{-3}$ lb/$in.^2$ = 47.88 $N/m^2$

Table 3. Energy

1 J = .2389 gm cal = $9.48 \times 10^{-4}$ BTU = $2.778 \times 10^{-4}$ W-hr

1 BTU = $1.0548 \times 10^3$ J = .293 W-hr = 252 gm cal

1 gm cal = 4.186 J = $3.968 \times 10^{-3}$ BUT = $1.163 \times 10^{-3}$ W-hr

1 W-hr = 3.413 BTU = $3.6 \times 10^3$ J = 860 gm cal

1 W-yr = $3.156 \times 10^7$ J = $2.992 \times 10^4$ BTU

1 ft-lb = 1.356 J = $1.285 \times 10^{-3}$ BTU

1 MeV = $1.602 \times 10^{-13}$ J

1 AMU ↔ 931.5 MeV = $1.492 \times 10^{-10}$ J

1 hp-hr = $2.685 \times 10^6$ J = 2545 BTU

1 lb coal = $1.3 \times 10^4$ BTU = $1.4 \times 10^7$ J (average energy content)

1 bbl oil = $5.9 \times 10^6$ BTU = $6.2 \times 10^9$ J (average energy content)

1 $ft^3$ natural gas = 1000 BTU = $1.05 \times 10^6$ J (average energy content)

Table 4. Power

1 W = .2389 gm cal/sec = $9.480 \times 10^{-4}$ BTU/sec
= 3.413 BTU/hr = 81.97 BTU/day = $2.992 \times 10^{4}$ BTU/yr
= 8766 W-hr/yr = $3.15 \times 10^{7}$ J/yr

1 gm cal/sec = 4.186 W

1 BTU/sec = $1.0548 \times 10^{3}$ W

1 BTU/hr = .293 W

1 BTU/day = .0122 W

1 BTU/yr = $3.34 \times 10^{-5}$ W

1 W-hr/day = .0417 W

1 W-hr/yr = $1.141 \times 10^{-4}$ W

1 hp = 550 ft-lb/sec = 745.2 W

Table 5. Useful Information

Total surface area of the earth = $5.10 \times 10^{14}$ $m^2$

Total land area of the earth = $5.79 \times 10^{7}$ $mi^2$
= $1.5 \times 10^{14}$ $m^2$

U.S. land area = $3.6 \times 10^{6}$ $mi^2$ = $9.3 \times 10^{12}$ $m^2$

Radius of earth (equator) = 3963 mi = $6.378 \times 10^{6}$ m

Volume of oceans = $1.37 \times 10^{18}$ $m^3$

Mass of atmosphere = $5.14 \times 10^{18}$ kg

Average distance to sun = $1.495 \times 10^{11}$ m

Sun's output = $3.9 \times 10^{26}$ W

Sun's average energy flux at the top of the atmosphere = 1400 $W/m^2$

1 year = $3.156 \times 10^{7}$ sec

1 day = 86400 sec

Density of air at STP = 1.293 $kg/m^3$

Speed of sound 20° C, dry = 343.4 m/sec = 1126.6 ft/sec

U. S. population = 207 951 000 (May 1972)

World population = $3.85 \times 10^{9}$ (1973)

U.S. total power = $6.6 \times 10^{19}$ J/yr = $2.1 \times 10^{12}$ W (1970)

World total power = $2 \times 10^{20}$ J/yr = $6.4 \times 10^{12}$ W (1970)

Steffan-Boltzmann constant $\sigma$ = $5.67 \times 10^{-8}$ $W/m^2$ $°K^4$

Wien's law constant = 2897 micron-degrees

hc = 12400 eV - Å

1 Curie = $3.7 \times 10^{10}$ decays/sec

1 Roentgen = $3.336 \times 10^{-10}$ $Coulombs/cm^3$

1 rad = $10^{-2}$ J/kg

Avagadro's number = $6.02 \times 10^{23}$ molecules/gm mole

ppm = parts per million, usually on a weight basis for solids and liquids and on a volume or molecular basis for gases.

# Appendix 5
## Table of powers of 2

| n | $2^n$ | $2^{-n}$ |
|---|---|---|
| 0 | 1 | 1 |
| 1 | 2 | .5 |
| 2 | 4 | .25 |
| 3 | 8 | .125 |
| 4 | 16 | .0625 |
| 5 | 32 | .03125 |
| 6 | 64 | $1.56 \times 10^{-2}$ |
| 7 | 128 | $7.81 \times 10^{-3}$ |
| 8 | 256 | $3.91 \times 10^{-3}$ |
| 9 | 512 | $1.95 \times 10^{-3}$ |
| 10 | 1024 | $9.77 \times 10^{-4}$ |
| 11 | 2048 | $4.88 \times 10^{-4}$ |
| 12 | 4096 | $2.44 \times 10^{-4}$ |
| 13 | 8192 | $1.22 \times 10^{-4}$ |
| 14 | 16384 | $6.10 \times 10^{-5}$ |
| 15 | 32768 | $3.05 \times 10^{-5}$ |
| 16 | 65536 | $1.53 \times 10^{-5}$ |
| 17 | 131072 | $7.63 \times 10^{-6}$ |
| 18 | 262144 | $3.81 \times 10^{-6}$ |
| 19 | 524288 | $1.91 \times 10^{-6}$ |
| 20 | 1048576 | $9.54 \times 10^{-7}$ |

$$2^n \simeq 10^{.3n}$$